A PHILOSOPHER READS...

Marvel Comics'

Daredevil

From the Beginning to *Born Again*

Marvel Comics'

Daredevil

From the Beginning to *Born Again*

Mark D. White

Ockham
Publishing

Published by Ockham Publishing in the United Kingdom in 2023

ISBN 978-1-83919-571-6

Cover by Claire Wood

www.ockham-publishing.com

Acknowledgments

I would like to thank Rob Johnson and everyone at Ockham for their ongoing faith in this series.

I thank Christine Hanefalk, whose love and passion for Daredevil has been an inspiration for many years and whose comments on an early draft proved invaluable.

I thank my good friends Lauren Hale, Carol Borden, Anita Leirfall, and Bill Irwin for their endless support.

Finally, I thank all the creators who have chronicled Matt Murdock's adventures over the years, especially Stan Lee, Bill Everett, and Wally Wood, who created him, and Frank Miller, Klaus Janson, David Mazzucchelli, and John Romita, Jr., who perfected him (in all his imperfection).

Contents

Introduction

I don't know about you, but I've always envied people who seemed to "have it all together." We all know people like this, who somehow manage to integrate all the various aspects of their life, both external and internal, into a tidy, cohesive package. Of course, we can never know how smooth and successful their lives truly are; their outward appearance may be managed and polished like an Instagram influencer's, while behind the curtain (or screen), their actual lives are a mess.

Although some people manage their lives better than others, we all have complications that we struggle to reconcile. Even if we can balance the many external roles we fill—friend, family member, romantic partner, employee, community member—we also need to balance our internal lives, including our desires, impulses, and moral code, all of which may be multifaceted.

This process of "getting your life together" can be difficult for anybody, and most of us have struggled with this to some extent at some point in our lives. Unfortunately, some people find it particularly difficult, especially with respect to their thoughts and feelings; such people may cycle through good times and bad times their entire life, some gradually improving over time, and others spiraling downwards into ever worse states of internal disarray.

This is the point at which I introduce the subject of this book, a man who may appear on the outside to have made a successful life for himself, but on the inside, he "contains multitudes," as Walt Whitman famously wrote in *Leaves of Grass*.

Matthew Murdock is, famously, a man of contrasts. He was an only child, raised in poverty and growing up to become a successful lawyer—largely due to his single father, a boxer and part-time mob enforcer who pushed his son to study rather than fight. Matt was blinded by a freak accident that also heightened his other senses, granting him a range and depth of perception that sighted people can only dream of. He uses these senses, together with fighting skills acquired in violation of his father's wishes, to fight crime as the costumed vigilante Daredevil—and, in the process, regularly violates laws he has sworn to uphold as a member of the bar.

Beneath the surface, we discover even more fascinating conflicts. He is perpetually unsettled about who he is or should be: Matt Murdock or Daredevil, man or hero, lawyer or vigilante. He is obsessively concerned with whether he makes a positive contribution as Daredevil, or does "good" for the world, and assuming he does, he questions whether that good is his true motivation for doing it. He places a high value on rules but repeatedly finds himself in the position of breaking them—and even when he can justify doing this, he worries about the justification itself, and what it means that he had to find a justification at all. (Exhausting!) Most superheroes struggle from time to time with who they are and what they're doing, but none as unrelentingly as Daredevil, and as soon as he comes to some resolution, something inevitably happens to make him doubt himself once again.

All of these conflicts help to create one of the most dramatic characters in superhero comics. This is unfortunate for Matt Murdock, of course, but it's great for the creators of his comics, who rank among the best in the industry, attracted to the character because of his complexity: outwardly charming yet inwardly tortured, one moment ecstatic and the next despondent. It's also great for his fans and gives us a chance to explore the philosophy behind Matt's internal conflicts, as well as what they mean for the man who suffers them.

In this book, I'll focus on the early years of Matt Murdock and Daredevil, from his first appearance in 1964's *Daredevil* #1 to 1986's *Daredevil* #233. The latter was the final issue of the classic "Born Again" storyline, which also marked the end of legendary comics creator Frank Miller's time on the book. (I'll also discuss the 1993-1994 miniseries *The Man Without Fear*, in which Miller expands on Daredevil's origin with many of the concepts he introduced during his run on the monthly title.) As is true with many long-running comic book characters, the Matt Murdock we know today is not the Matt Murdock we met in 1964, but it is fairly close to the one that Frank Miller wrote. Although Miller is properly credited for crafting the modern version of Daredevil, he didn't radically change or "reboot" the character. Rather, he refined and emphasized certain aspects of Matt Murdock that had been gradually introduced by his creative teams since the beginning. Miller made the characterization of Matt Murdock more consistent, which ironically allowed his internal contracts and conflicts to come to the surface more clearly— and dominate most of his stories ever since.

Through this short book, we'll explore the moral character and behavior of Matt Murdock, including his belief in the importance of rules, the role of promises in his life, and his sense of responsibility, each of which he tends to take farther than he should. We'll also see how his penchant for violence grew over the years, as well as how far he'll take it—and why. Then we'll look more deeply into who Matt Murdock is— why he does the things he does, in terms of both motivation and the basic character traits that influence his choices (for better or worse). Finally, we'll see how he gradually starts to come apart at the seams, the roots of which appear early on but intensify as the years go on, reaching a climax in "Born Again" as his personal dissolution gets a little push from someone else.

Matt's recovery from this final episode suggests ways for us to deal with the myriad internal conflicts we might have in our lives (even if they are less dramatic and colorful than Daredevil's). Future stories in the comics—and possibly future books in this series—will reveal whether Matt manages to hold himself together in the long run. For now, though, we'll learn how Matt Murdock became the Man without Fear we've come to know and love, and who continues to thrill us to this day in the comics, movies, and television series.

Chapter 1: Rules, Law, and Justice

The Mandatory Origin Scene

The story of Matt Murdock and how he became Daredevil is simple and straightforward, although significant additions were made to it over the years.[1] Matt was raised in Manhattan's "Hell's Kitchen" neighborhood by his father Jack, an alcoholic boxer (and part-time enforcer for the mob) known professionally as "Battlin' Jack Murdock." (Nothing is revealed about Matt's mother until much later.) Jack desperately wanted his son to grow up to have a better life than him, so he forced his son to spend all of his free time studying, leaving little time for play. This led to brutal teasing from the other kids, but Jack forbade Matt from fighting, which only frustrated Matt further, fueling the rebellious streak that led him to train in private on his father's punching bag.

Young Matt's life changed forever when he ran in front of a truck to save an elderly man in its way. After the radioactive waste in the back of the truck spilled and splashed into his eyes, Matt lost his sight, but his other senses were heightened tremendously. Together they endowed Matt with a "radar sense" that gave him the ability to perceive surfaces, 360 degrees in every direction around him.[2] Soon afterward, a mysterious blind man named Stick found Matt and trained him to use his enhanced senses, not only to navigate his new world but also to help him fight in various styles and perform amazing feats of athletic and acrobatic skill, satisfying his desire for action and danger that predated his accident (but which was held back by his father).

As a teenager, Matt attended one of his father's matches, unaware that Jack was ordered by his mafia boss to throw the fight. However, during the match, Jack realized he had a chance to win, and after seeing Matt proudly cheering him on, he did exactly that—after which his boss had him murdered in the alley outside the arena. Devastated and enraged, Matt went after the men responsible, using the skills Stick gave him to exact revenge, beating them savagely (but stopping short of killing them). Even if he wasn't yet in costume or using a codename—the accounts differ on these points—this was the first time he used his abilities and training to strike back against crime, and can reasonably be considered the point at which Daredevil was born.

"Battlin' Jack" Murdock

Appropriately enough, Jack Murdock gives us a good way to start exploring his son's complicated relationships with rules. Like his grown son, Jack had two lives as well: one of them legitimate, as a boxer, and the other of them criminal, as an enforcer for the mob, both of which have analogues to Matt's later life as a lawyer and a superhero. Although boxing may seem to have little to do with the practice of law, they share a basic reliance on a system of rules. Sports, like games in general, are defined by the rules that make them up. What is football or chess or poker, after all, but a system of rules dictating the order of play, the permissible moves, and the definition of victory? Just consider that whenever significant changes to a sport's rules are suggested, there is a predictable outcry from fans that "this will change the sport" or "this is no longer the same sport!"[3]

In addition to the rules that define the parameters of a game, we also need rules that spell out the consequences for breaking them. This is important because one player cannot be allowed to make exceptions to the rules for themselves while all the other players continue to follow them, and also because, if all the players are not following the same rules, they're literally not playing the same game.

When Jack agrees to throw matches on the orders of his mob boss, this contradicts the spirit and purpose of the game: seeing two trained fighters try their best to defeat each other within rules that define fair competition. This also provides entertainment to the fans and allows gambling odds to be determined on an honest assessment of the skills of the respective boxers. Of course, it was the latter purpose that Jack's boss meant to exploit by secretly influencing the outcome of the match and taking advantage of the honest odds—an end that was frustrated by Jack's honest victory in what would be his final match.

Although Jack followed the rules of boxing in most of his matches, he routinely broke society's rules—better known as laws—when he threatened people who owed money to his boss. In general, laws can be understood as the rules society establishes to govern interactions between people in order to minimize conflicts, protect rights, and ensure justice (among other goals the law might promote). Again, Jack's boss takes advantage of the fact that most people follow the law by breaking it himself, creating and enforcing his own rules in the shadow of the official ones— and trusting that his "business partners," who are also complicit in this illegal activity, will be unlikely to report it to the authorities.

Ironically, it was Jack's refusal to break both the rules of boxing and the laws against fraud in his final match that resulted in his death, which set Matt on his path to fighting crime as Daredevil. However, Jack didn't fight honestly out of a devotion to fair competition for its own sake; rather, he decided during the bout itself that he wanted to set a good example for his son. As he tells his opponent while he pummels him, it was "my one chance…maybe my last chance…to make Matt proud of me," showing him that it was important to do the right thing regardless of the consequences.[4] In contrast to Jack's paternal reason for following the rules in this case, Matt would come to develop a personal embrace of rules for their sake, which makes his future violation of them represent even more of a deep conflict within his own identity.

Rules Help People Be Better

Matt's relationship with rules stems from an incident when he was a young boy. Even though he had promised his father he would never fight, one day Matt got into a fight after school, breaking a rule he agreed to follow. When Matt told his father what happened, Jack hit his son in a moment of drunken rage. Although Jack immediately realized what he had done and apologized, it made an indelible impression on the boy, who later remembered:

> I spent the night perched on the Brooklyn Bridge, thinking...thinking about right and wrong...and how even my father could do bad things...how even Dad needed rules to obey. Rules...laws. My morning, the course of my professional life was set. I would become an attorney.[5]

In one retelling of this story, the narration concludes that "the only way to stop people from being bad is to make rules. Laws....He will study the rules. He will study the laws."[6] Here, we see that Matt acknowledges that people are imperfect, often tempted by inclination or emotion to behave unethically, and that rules help people fight these influences—and hold them accountable if they fail. This reflects a focus on the importance of rules to the individual, in the form of either moral or legal rules, rather than to society as a whole.

Many philosophies and religious traditions recognize that human beings are imperfect or fallible. Although it is not mentioned as much in the early comics as it will be later, Matt's own faith, Catholicism, maintains that humans are "fallen" and carry the stain of original sin through their lives, making them subject to temptation and lapses in morality, which should be confessed and repented. Philosophers who follow *virtue ethics*, a moral philosophy that we'll discuss in more length later, also believe that people don't always make the best decisions, even if they have good intentions and are generally ethical (or virtuous) in nature.

Perhaps the most prominent secular philosopher to emphasize the moral imperfection of human beings and the importance of rules in helping them make good choices is Immanuel Kant, the 18th-century philosopher best known for his *categorical imperative*, a formal ethical test that results in *duties* that forbid unethical behavior. Although his own relationship with Catholicism was fraught at best, Kant wrote that only divine beings had perfect wills, whereas human beings are necessarily imperfect, often tempted by desires, corrupted by anger and jealousy, and unduly influenced by other people.[7] To effectively resist these factors, we must commit ourselves to observing the moral law as embodied in the categorical imperative and following the duties it issues. These duties can be understood as rules, which according to Kant must be followed for their own sake, rather than some other advantage one can gain from being ethical, in order to make a person truly moral.[8] (This will be important when we discuss Matt's motivation to be Daredevil later.)

Of course, Matt realizes he is as imperfect as anyone, and rules are just as necessary to keep his own impulses in check as anyone else's. This becomes apparent after he meets Elektra Natchios, an alluring fellow student at Columbia University. The first thing he notices about her is her scent as he is running and jumping over rooftops at night; he pursues her to Central Park, where he finds only her clothes, which leave "a trail. Quite a trail." Before he has a chance to react, two police officers arrive, asking about the woman who left the clothes, and Matt considers escaping. "He could flatten the cops. It would be easy," reads the narration. "But that would be breaking the rules. That would be breaking the law."[9] Here, Matt exemplifies the Kantian ideal of following the rules because they are rules, rather than to avoid punishment or other negative consequences for breaking them.

This is not to say that Matt does not appreciate these less noble aspects of rules as well. When Matt begins to suspect Elektra is more than she seems, he sneaks around her father's mansion during a party to learn more, but his attempts at stealth are thwarted when he kicks an armed guard through a window—and then through a skylight over the ballroom where the party is being held. As more guards shoot at him, Matt again curses himself for breaking the rules and making "a mess of things," which suggests a self-interested motivation more than an ethical one.[10]

Rules can also help us avoid bad decisions when we're upset and not thinking "straight." When considering resisting arrest in the park, Matt acknowledges his own rash nature when he remembers when he accidentally knocked a woman through a window to her death while angrily pursuing his father's killers.[11] His horror at the thought of losing control like that again represents an awareness that he was in what psychologists call a "hot state," in which our emotions overwhelm our rational faculties. This realization gives him even more reason to follow rules and the laws, and so he surrenders to the police—who let him go anyway after realizing he's blind.[12]

One of the "bad decisions" Matt made in an overly emotional state may have been to stay with Elektra, even after both his mentor Stick and his college roommate Foggy Nelson urge him to stay away from her—*and* after she admits to him how many people she's killed. After her father is murdered, it is Elektra who leaves Matt, who blames himself for the pain he feels, considering it the punishment for "letting his wild part run free. For breaking the rules."[13] This could be an early glimpse of divine retribution, but more likely Matt considers his heartache to be simply one of the negative consequences of bad decision-making that rules—not to mention common sense—are designed to prevent.

Rules Also Hold Society Together

In the last section, we focused on the importance of rules to a person's own decision-making, but they also have an essential role to play in society, specifically in the form of laws—a subject Matt knows very well.

In an early story, Matt gives a guest lecture at the fictional Carter College, and he begins with the statement that "only the law stands between justice and total anarchy," a natural extension of his reflection on rules after his father hit him as a boy.[14] The political philosopher Thomas Hobbes showed the same flair for language when he wrote that when "men live without other security than what their own strength and their own invention shall furnish them," their life will be "solitary, poor, nasty, brutish, and short."[15] Hobbes was describing the "state of nature" from which we can escape to some measure of security only by forming a government of laws and rules to protect citizens from each other. Such rules naturally begin with laws against violence and theft, thereby protecting persons and property, and then develop to protect a wider range of rights and interests, with some laws enforced by the state (criminal laws) and others enforced by affected individuals (civil law).

As a society's system of laws grows, it moves farther from anarchy, as Matt puts it, and closer to justice. Over 2500 years ago in the *Nicomachean Ethics*, the philosopher Aristotle wrote of several types of justice, including *distributive justice*, which governs how resources are shared or assigned; *corrective justice*, which deals with addressing harms

11

and wrongs to individuals; and *retributive justice,* which ensures that wrongdoers are punished appropriately for their crimes.[16] Although the last two are closest to what Matt was talking about, all of these specific conceptions of justice boil down, in one way or another, to fairness, making sure that all people get what they are owed, whether in terms of resources, attention, or simple respect. In fact, Matt's follow-up statement at Carter College emphasizes the "all" in the last sentence: "Law must offer equitable protection to all—regardless of race, creed, or color."[17]

In a later story, after a woman asks Matt who he works for, he answers, "I don't. I serve an idea, an ideal, a concept…something called justice."[18] This may seem like an odd thing for attorney Matt Murdock to say but a more natural one for Daredevil, given the way he straddles both sides of the law that Matt is sworn to uphold. But there is no contradiction if we understand the relationship between law and justice: Justice is an abstract ideal that laws, as products of imperfect human minds operating in particular contexts and with various incentives, can only approximate.

For example, in one story, Daredevil stops a young boy named Billy from shooting Hogman, a career criminal he saw murder someone before poisoning Billy's coach. After Matt hands Hogman over to the authorities, Billy asks him if he can promise the killer will see justice, which gives Matt a chance to explain the importance, as well as the limitations, of rules and laws:

> We're only human, Billy. We can be weak. We can be evil. The only way to stop us from killing each other is to make rules. Laws. And stick to them. They don't always work, but mostly they do. And they're all we've got.[19]

It is common for superheroes, who often operate on both sides of the law, to emphasize the difference between law and justice as they violate the former to serve the latter, but Matt's role as a lawyer puts him in a unique position to appreciate this trade-off, which contributes to the internal torment that contributes to his being such a fascinating character.

This story provides another example of how laws promote justice, for Billy is not the only one who almost kills Hogman that day—Frank Castle, otherwise known as the Punisher, is also gunning for him (literally). Although Daredevil and the Punisher both feel they serve justice, Castle's approach is more definitive and final, taking it upon himself to decide who is deserving of punishment and then meting it out, usually fatally. Although Castle and Billy do not have much in common, they can both be said to be acting out of vengeance, a personal drive to punish those who have hurt them, rather than justice, or to see an abstract ideal of balance and fairness restored. Vengeance is a natural impulse, but society constrains it by enforcing laws that reserve the right to punish wrongdoers for the state. The state—in the form of courts—does so in a setting that ideally reduces the case to facts, logic, and reason, so when wrongdoers are convicted and punished, they cannot reasonably argue simply that "someone was out to get me."[20]

Of course, Matt himself is no stranger to vengeful impulses, as shown by the way he beat the men responsible for his father's death. After that uniquely personal episode, his respect for rules and law mostly keeps these emotions in check. For example, one of Daredevil's many battles with the psychotic killer Bullseye ends with the latter unconscious on a subway track. When Daredevil hears a train approaching, he starts making excuses for leaving Bullseye there, thinking to himself, "Nothing I can do...too weak...I'm too weak to lift him," and then actually justifying it: "You deserve to die, Bullseye...you'd just kill again...I hate you..."[21]

13

In the end, though, Daredevil pulls Bullseye off the track, only to face New York City police lieutenant Nick Manolis, who argues that he should have left him to die because Bullseye will now likely be acquitted at trial due to a brain tumor that he could blame for his crimes. As he did with Billy, Matt explains to Manolis why it is not up to either of them to decide whether Bullseye is punished with death:

> Nick, men like Bullseye would rule the world—were it not for a structure of laws that society has created to keep such men in check. The moment one man takes another man's life in his own hands, he is rejecting the law—and working to destroy that structure. If Bullseye is a menace to society, it is society that must make him pay the price. Not you. And not me.

As he continues, Matt is upfront with Manolis about his feelings regarding Bullseye, adding a religious dimension to his opposition to killing: "I wanted him to die, Nick. I detest what he does…what he is. But I'm not God—I'm not the law—and I'm not a murderer." Manolis points out that Bullseye will surely kill again, and says that it will be Matt's fault—and even though Matt knows, rationally, that that is not true, it's something he feels all too deeply in his heart, as we shall see when we discuss his refusal to kill at more length later.

Matt's conflicted relationship with Elektra highlights another wrinkle in his attitude toward laws and justice. As opposed to Bullseye, whom Matt would like to kill out of vengeance, Elektra is a killer Matt feels great affection for, and therefore he is resistant to apprehend her at all (much less punish her). When Matt encounters her for the first time since college and realizes she has become a paid assassin—which is "everything I despise"—he remembers their time together and the ordeals she suffered that steered her to her current path. Then he thinks to himself, "But it doesn't count. None of it. No matter how much it pains me,

I must hunt Elektra down… and bring her to justice!"[22] Nonetheless, the next time Matt confronts her, together they defeat a gunman trying to kill both of them, after which he takes the gunman in, not Elektra—showing that he can become morally conflicted about the people he loves as well as the people he hates.[23]

Idealistic as he can seem, Matt occasionally gets despondent about the ability of the law to even approximate the cause of justice. In one tale, after a criminal escapes his grasp, he reflects on how cynical he's become about humanity—"Lately, every case I've been involved in just convinces me more and more how sick we are"—as well as the law itself:

> Even the one thing I always had real faith in—the law—even the law seems to work backwards these days. And Matt Murdock has devoted his life to preserving the sanctity of the law, the system of due process, the ideal of equal justice for all. I'm beginning to feel very alone in that devotion.[24]

Perhaps Matt's growing doubts about the law result from the conflict between the rules he tries to follow for their own sake and the better outcomes that following the rules often prevents—not better outcomes for himself, but in support of other values and principles he holds dear.

[1] The basic origin was first told in *Daredevil* #1 (April 1964) and expanded in *Daredevil* #53 (June 1969), but many of the details given here were added much later, including the background Matt gives reporter Ben Urich in *Daredevil* #164 (May 1980) and the expanded origin presented in the 1993-1994 miniseries *Daredevil: The Man Without Fear*, which has become canonical.

[2] For an exploration of the science behind Matt's enhanced abilities, including his radar sense, see Christine Hanefalk's definitive book *Being Matt Murdock: One Fan's Journey into the Science of Daredevil* (TOMP Press, 2022).

[3] For philosophical reflection on sport, see David Papineau, *Knowing the Score: What Sports Can Teach Us about Philosophy (and What Philosophy Can Teach Us about Sports)* (New York: Basic Books, 2017). (The chapters in Part II discuss rules specifically.)

[4] *Daredevil* #164 (May 1980). If we wanted to go deeper into the ethics of Jack's decision, we could question his decision on the grounds that, by accepting death, he would be deserting his young, newly blinded son when he needed a parent the most. (In the earliest renderings of this event, at least, Matt is drawn as a pre-teen, who somehow became older in later versions!) I thank Christine Hanefalk for suggesting this point.

[5] *Daredevil* #191 (February 1983).

[6] *Daredevil: The Man Without Fear* #1 (October 1993).

[7] Immanuel Kant, *Grounding for the Metaphysics of Morals*, trans. James W. Ellington (Indianapolis, IN; Hackett Publishing, 1785/1993), especially p. 414.

[8] A good place to start learning about Kant, duties, and the categorical imperative is Roger J. Sullivan, *An Introduction to Kant's Ethics* (Cambridge: Cambridge University Press, 1994).

[9] *Daredevil: The Man Without Fear* #2 (November 1993).

[10] *Daredevil: The Man Without Fear* #3 (December 1993).

[11] *Daredevil: The Man Without Fear* #2 (November 1993).

[12] For more on the dangers of making choices while in a "hot state," as well as other issues with compromised rationality, see Daniel Kahneman, *Thinking Fast and Slow* (New York: Farrar, Straus and Giroux, 2013).

[13] Ibid.

[14] *Daredevil* #28 (May 1967). In case anyone reading this book still doubts the relevance of superhero comics to philosophy or vice versa, the credits in this issue read: "Philosophically Produced by: Smilin' Stan Lee and Genial Gene Colan." (So there.)

[15] Thomas Hobbes, *Leviathan* (Indianapolis: Hackett Publishing, 1651/1994), p. 76.

[16] Aristotle, *Nicomachean Ethics*, trans. Terence Irwin (Indianapolis, IN: Hackett Publishing, 1985), particularly Book V.

[17] *Daredevil* #28.

[18] *Daredevil* #199 (October 1983). Earlier, when Ben Urich asks Matt why he became Daredevil, he says, "I think you already know the answer to that, Ben. Justice. Blind justice" (*Daredevil* #164, May 1980).

[19] *Daredevil* #184 (July 1982).

[20] We'll see Frank again when we discuss killing in chapter 5.

[21] *Daredevil* #169 (March 1981). (All quotes from this paragraph and the next are from this issue.)

[22] *Daredevil* #168 (January 1981).

[23] Another time, to his credit, Matt does tell her, after they defeat dozens of ninja, "You're going to jail, Elektra. You're a killer...a cold-blooded assassin...and I'm taking... you... in," but collapses from exhaustion before he can act on it (*Daredevil* #175, October 1981).

[24] *Daredevil* #115 (November 1974).

Chapter 2: Ethics, Promises, and Judgment

The Basic Ethics of a Daredevil

How can we best describe Matt Murdock's ethics? Based on his personal devotion to rules, our first thought may be that Matt is a deontologist. *Deontology* is the school of moral philosophy that judges actions by holding them up to some principle or ideal. A deontologist may say that lying is wrong because it fails to respect the person being lied to, or that killing is wrong because life is considered sacred; in both cases, we judge the act according to a more general principle that the deontologist regards as important. Note that we don't ask what happened as a consequence of a particular instance of lying or killing; the acts themselves are considered wrong to a deontologist based on a principled judgment of them, regardless of their outcomes.[1]

Because of their foundational belief in principles and ideals, deontologists speak of actions in absolute terms like "right" or "wrong" rather than comparative terms like "good and "bad," which are used by consequentialists. *Consequentialism* is the ethical system most often contrasted with deontology, and it judges actions based on the outcomes they lead to. There are many ways to define and evaluate the consequences of actions, and as a result, there are many varieties of consequentialism that differ according to how they define "better" or "worse" outcomes. The most well-known version of consequentialism is *utilitarianism*, which evaluates actions in terms of the resulting changes in "utility," which is usually understood to mean happiness or well-being. To use the same examples as before, utilitarians would judge lying to be bad in general because it usually results in somebody getting hurt either by the deception itself or when it's found out; the same goes for killing, due to the obvious harm to the person killed as well as to their loved ones.[2]

In its strictest form, rule-following would seem to be insensitive to consequences, so it is easy to associate such behavior with deontology. However, utilitarians can endorse rules too. You no doubt noticed that, in the brief discussion of utilitarianism above, I often used the phrase "in general." Although lying is usually harmful, especially when done for selfish reasons, specific instances of it could be beneficial, such as "white lies" told to protect someone's feelings. By the same token, killing is usually bad—very, very bad—but taking a life in self-defense or in defense of another can be good (assuming there was no other option).

This last exception may be fairly clear, but in most cases of potentially harmful behavior, the outcomes are more difficult to evaluate, especially in the moment. This points out a practical problem with utilitarianism: calculation, or identifying the positive and negative outcomes of decisions, determining their magnitude, and comparing them to arrive at an overall assessment. Because of this difficulty, some philosophers advocate for *rule utilitarianism*, which recommends that people follow rules, such as "do not lie" and "do not kill," that are designed to generate good outcomes in most cases, even if they will occasionally prevent specific actions that are beneficial on the whole.[3]

So, where does Matt fall? As we have seen, he believes in rules mainly because they help control people's worst impulses (including his own) and therefore protect others from their wrongful actions. Stated this way, we see that Matt's general ethical behavior is supported by *both* deontology and utilitarianism: Preventing harm supports the ideal of justice (specifically corrective justice, in a pre-emptive sense) and also the utilitarian goal of increased well-being. Both reasons may account for Matt's general orientation toward following rules and laws, but they do so in slightly different ways. This becomes problematic when a rule supports doing something that may lead to bad consequences, such as saving Bullseye's life even though it will likely result in his killing more innocent people in the future. Like most heroes, Matt has both a deontological side that pushes him to "do right" and a utilitarian side that pushes him to "do good"—directives that are aligned in most cases, but when they do not, it can generate intense moral conflict.[4]

Promises Made

As we begin to look at specific aspects of Matt Murdock's ethical behavior, we'll start with a simple and ordinary example: promises. A promise is a unique type of rule because it's one we create for ourselves: We promise someone else we will do something (or abstain from doing something), and by doing so we set a rule for ourselves. In general, keeping a promise is supported by both deontologists and utilitarians: It is the right thing to do because it honors the commitment one makes to another person and treats them with an appropriate degree of respect, and it is a good thing because, narrowly speaking, it prevents the person the promise was made to from being disappointed, and more broadly it increases the range of human cooperation (in the same way that contracts enable more business transactions).

As we would expect from someone devoted to rules, Matt takes promises very seriously. The most important promise in his life was the one he made to his father to study hard and avoid fighting, so he can make more of himself than his father did. In the version of his origin he tells reporter Ben Urich (after he figures out Daredevil's secret identity), Matt remembers being a young boy and wanting to play outside with the other kids, but his father made him stay inside and study, saying,

I promised your momma before she...before she died...that I wouldn't let you grow up to be an uneducated pug like me. Now you have to promise me something, son. Promise me you'll study every chance you get, that you'll become a doctor or a lawyer...somebody important! Promise me that you'll be somebody I never could...

Matt agreed, saying, "I promise, Pop! You'll be proud of me! You'll see...," thinking to himself later, "I can't go against his wishes—not after all his sacrifices. I've got to be the son he wants me to be!"[5]

This crucial scene in the formation of both Matt Murdock and Daredevil contains many layers. (We'll ignore Jack's unfairly poor opinion of himself, which Matt does not share, the pivotal incident of violence notwithstanding.) For one, we note that Matt's promise to his father may not be completely voluntary; even if Jack does not literally force his son to make the promise, he does exert considerable pressure on the boy, dropping his mother's memory into his speech for added effect.[6] Also, I'm sure adult Matt would point out that, legally speaking, young Matt is too young to be competent to make a binding promise, further compromising any presumption that the promise was voluntary. The aspect of his age aside, Matt does enthusiastically agree to the promise, thereby making the commitment his own, at least in his own mind. In philosophical terms, by making that promise Matt has chosen freely to bind his will to the rule he made, a paradoxical act of personal freedom that limits his freedom at the same time.

Jack doesn't merely mention Matt's mother, though: He actually cites his earlier promise to her to make sure Matt turned out better than he did.[7] This motivates Jack's action, showing that he also takes promises seriously, and also adds weight to his request of Matt, extending his promise to his father to his mother as well. This aspect is emphasized in a later recounting of this scene, in which Jack says, "*We* owe it to your

21

mother, Matt. You've got to be something special. You've got to be nothing like me."[8] Furthermore, although he was unaware of who it was at the time, Matt also made a promise to his mother after the accident that robbed him of his sight: When he realized his other senses were heightened, a nun caring for him told him that "this…may not be a bad thing. What you could do with it…Just think of it as a blessing, Matt. It's yours. Yours. And it's our secret. Don't tell anyone. Promise me now…"[9] Although he did keep the secret from his father, he revealed it to many others over the years, mostly close friends and love interests, but interestingly not with many of his fellow superheroes (many of whom do not know he's even blind).

The fact that both Jack and Matt take their promises to Matt's mother so seriously, even though Matt (and possibly Jack) believed her to be dead, brings up the fascinating issue of the obligation we have to keep promises to the dead—which also applies to Matt's continued adherence to his promise to his father even after he is killed. In a later story, when Matt's girlfriend Heather Glenn, who was aware of his dual life, accuses him of neglecting his personal life (and her) in favor of being Daredevil, he thinks to himself:

> I don't live under the shadow of Daredevil. If anything, I live under the shadow of the promise I made to my father years ago. I swore to him I'd make something of myself, and I think I've suc-ceeded…both as Matt Murdock…and as Daredevil…[10]

The fact that both Matt and Jack continue to hold themselves to promises made to people even after their death shows that promises are not just about the people we make promises to. Although we speak of honoring the memories of the dead—which does take on an extra di-mension for those, like Catholics, who believe in an afterlife—this alone does not explain why we continue to keep promises that may involve costs to the promisor with no possibility of benefiting the promisee.[11] Keeping our promises to the dead is often more about the commitments we make ourselves, and *to* ourselves, for reasons having to do with more than just satisfying the other person's desires (while they were alive). This is especially true when the promises we make to others are motivated by their concern for us, such as when a good friend asks you to promise to quit smoking, finish a college degree, or write that screenplay you're always talking about. Even if that friend dies, her wishes still live on in you. Matt's promise was not just to benefit his father, but to do so through lifting himself up, which is still a valid concern even after his father's death.[12]

Promises Broken

Promises can be difficult to keep, regardless of whether the person the promise was made to is alive or dead. This shouldn't be surprising, because promises are often most necessary when it deals with something you don't want to do anyway: Jack made Matt promise to study (and not to fight) because he knew the boy wanted more than anything to play (and fight). Later, while training with Stick, Matt thought back to the promise he made to his father and said, "That promise made my childhood a misery, Stick. The neighborhood kids ridiculed me horribly, made me so angry…"[13] In a way, though, the burdens we are willing to endure to keep a promise demonstrate how much the promise means to us—and how much the person to whom we made the promise to means to us too. (For example, marriage vows wouldn't mean as much if they weren't at times hard to keep.)

This does not mean, however, that promises are absolute and that we never have valid cause to break them. We already saw that Matt broke his promise as a kid and got into a fight after school, after which his father hit him and instilled in Matt the importance of rules. Although he clearly resented the way his father reacted, Matt made no excuse for what he did himself—in that case, he realized he had no good reason for breaking his promise. This is true especially because what he did was in direct opposition to the reason behind the promise—to prevent him from becoming a common fighter like his father—rather than in support of another reason that might have more weight in a particular case (such as if he were defending a smaller child from bullies).

The more important way Matt broke his promise was that he regularly trained in secret in his father's gym. Years later, when revisiting the gym, Matt thinks about the promise he made to his father and then broke:

> You trained here. You trained and made me promise to stay away.
> So I lied to you and promised and came here anyway. I became a
> fighter. Just like you. Endless stolen hours on the ropes. At the bag.
> At the bag. On my mother's grave you made me promise.[14]

As Matt does above, we can draw a close link between promise-breaking and lying. Even if the broken promise was initially made sincerely, the act of breaking it later turns it into a lie after the fact, and like a lie, the broken promise uses the person the promise was made to as a means to the promisor's ends.[15] In this case, Matt used his father's trust in him, based on the promise he made, to gain the unsupervised freedom to access the gym and train in secret. Another way to look at it is that he continues to enjoy the benefits of the promise, such as his father's pride and admiration, while not bearing the costs (having to avoid anything to do with fighting).

Years later, while working with Stick to regain his lost radar sense, Matt confronts a vision of his late father, who calls him out on his broken promise and its equivalence to a lie:

> You promised me, Matt! You said you wouldn't fight! Look at you—you've already been in more fights than I ever was! You're gonna end up just like me...a worn out, used up old pug, with a face like a catcher's mitt and a brain made outta mush. You lied to me, son. On your mother's grave, you lied to me![16]

This time, though, Matt gave an explanation for breaking his promise, which he linked to his father's murder: "You just don't understand. It's not enough for me to just be a lawyer. The world needs Daredevil, Dad, to keep others from becoming victims like...like you..."[17]

As we'll see later when we explore Matt's motivations for being Daredevil, he regards both sides of his life, the lawyer and the superhero, as essential to fighting crime and protecting the innocent. As we see in the last quote, he feels that his quest for justice is enough to justify breaking his promise (such as defending another child might have justified his fighting when young). We can look at it as one principle, justice, conflicting with another principle, promise-keeping; both are important but Matt can only follow one, so he must use his judgment to determine which is more important. In general, we could say that once he realized that the law is only an imperfect approximation of justice, he also realized that fighting on both sides of the law was more important than keeping his promise to fight on only one.[18]

Judgment

The importance of judgment in choosing between two key moral principles is a regular aspect of any superhero's life, but none more so than

Daredevil, who embodies moral conflict like no other.[19] This is inevitable given Matt's dual missions to do right and to do good, which often coincide but create a unique inner torment when they diverge. This is especially true when he follows the rules or law but the result is bad for someone he cares about, for which he feels responsible even though he did the right thing (and the negative consequences were someone else's responsibility altogether).

This would not happen if Matt were strictly a deontologist, particularly a Kantian one. Kant was very adamant that doing the right thing, according to the moral law, was of paramount importance, and as long as you do the right thing, you're "in the clear" as far as the consequences are concerned. Kant took this to the extreme in his notorious "the murderer at the door" example. Imagine one day your best friend Foggy bangs at your door, tells you someone is trying to kill him, and asks you to hide him. You do so, and minutes later, the potential murderer, Bullseye, arrives at your door and asks if Foggy is there. What do you do? Kant infamously wrote that the duty not to lie matters above all else, so you have no choice but to tell Bullseye that Foggy is in fact there—and even if Bullseye murders your good friend, you can rest easy because you did the right thing.[20]

Even though Kant did write this—albeit in a more popular magazine of the day—this position is contradicted by his other more considered philosophical work, in which he very eloquently stresses the role of judgment in balancing conflicting obligations, such as telling the truth (to a psychopath) and saving a friend's life (from said psychopath, as it happens).[21] Nonetheless, the general point still holds: A deontologist would maintain that as long as you do the right thing, you are not morally responsible for any negative consequences (insomuch as they matter to a deontologist anyway). However, if you fail to do the right thing, you are responsible for any consequences. Both of these positions correspond to common-sense ethics: It is common to tell somebody who inadvertently makes a mess of things that "at least you did the right thing," which is presumed to make them feel better (and less responsible for the mess). By the same token, if we do something wrong, such as tell a selfish lie, then it is commonly believed that we are morally responsible for any negative consequences from it, such as anyone who gets hurt when the lie is discovered.

From a utilitarian perspective, however, it is only the consequences that matter, and there is no concept of doing the "right" or "wrong" thing aside from the good or bad outcomes of those actions. Rule utilitarians may be more conflicted if they follow a rule that is designed to generate good outcomes in most cases—their version of the "right thing"—but creates bad outcomes in a particular case. This better explains our intuitions with the murderer at the door: Even if we generally adhere to a rule of not lying, following that rule in this case would clearly result in horrible consequences, for which we would likely feel completely responsible regardless of the fact that we did the "right thing" by not lying. This tension is what you get when you introduce the deontological concept of rule-following to a consequentialist moral system, which may seem to complicate an otherwise straightforward idea, but it does provide us with a more realistic description of how many of us think about ethical issues. (No one ever said ethics were simple!)

This uneasy combination of deontology and utilitarianism describes Matt Murdock too. His devotion to rules is no stronger than his sense of responsibility for the consequences of his actions, even when he does the "right thing." Nowhere is this shown better than in his refusal to kill his most murderous foes even when he knows full well they will kill again if they have the opportunity. Earlier we discussed a fight between Daredevil and Bullseye in which Matt had a chance to let Bullseye be killed by a subway train if he just walked away, but he couldn't bring himself to let a man die when it was in his power to prevent it. When Daredevil next encounters Bullseye, he tells him, "just a few weeks ago I saved your life. I can't help but feel responsible for what you do with it."[22] Matt undoubtedly did the right thing, and this should relieve him of any responsibility for what comes next, but he does not believe this, at least in his heart.

What Bullseye does next is inevitable, of course: After accepting a contract to kill Wilson Fisk, also known as the Kingpin, he first kills one of his thugs. When Matt learns of this, he once again recalls saving Bullseye's life, and thinks to himself, "I've been wondering ever since how I would feel if you killed again! Now I know."[23] Soon he has a chance to tell the killer:

> Do you remember the subway, Bullseye? Do you remember how I stood over you, bleeding, barely conscious? I had beaten you. You were helpless, on the tracks, when a train approached. It was going to hit you.
>
> I wanted it to.
>
> I wanted to hear your bones splinter beneath its wheels. I wanted to hear you scream—and die. But I pulled you from the tracks. I saved your life. And you went free. And you killed—again.
>
> I've been carrying that around inside me ever since, Bullseye. It hurts. It hurts a lot. Now, I'm going to share that hurt with you.[24]

Matt's responsibility for Bullseye's actions may not be rational, but this passage shows that it's visceral, demonstrating how deeply his feeling of responsibility goes, even after he does the right thing—which, note, he never questions seriously, however much he regrets it.

[1] Immanuel Kant, discussed earlier, is widely considered the most important deontologist, but for more on deontology in general, see Larry Alexander and Michael Moore, "Deontological Ethics," *The Stanford Encyclopedia of Philosophy* (Winter 2021 Edition), Edward N. Zalta (ed.), at https://plato.stanford.edu/archives/win2021/entries/ethics-deontological/. *(The Stanford Encyclopedia of Philosophy is an invaluable—and free!—resource to which I will refer you often throughout this book.)*

[2] For more on consequentialism and utilitarianism, see Walter Sinnott-Armstrong, "Consequentialism," *The Stanford Encyclopedia of Philosophy* (Winter 2022 Edition), Edward N. Zalta & Uri Nodelman (eds.), at https://plato.stanford.edu/archives/win2022/entries/consequentialism/.

[3] For a thorough overview of rule utilitarianism, see Brad Hooker, "Rule Consequentialism", *The Stanford Encyclopedia of Philosophy* (Spring 2023 Edition), Edward N. Zalta & Uri Nodelman (eds.), at https://plato.stanford.edu/archives/spr2023/entries/consequentialism-rule/.

[4] Another prominent deontologist, W.D. Ross, emphasized this contrast in his book *The Right and the Good* (Oxford: Oxford University Press, 1930/2003).

[5] *Daredevil* #164 (May 1980).

[6] Matt would not learn his mother was alive for many years after this; it is revealed later that Jack was aware she was alive at the time, but told Matt she was dead to protect him from the fact that she left them (albeit for good reason, as Matt learned much later in *Daredevil*, vol. 4, #7, October 2014). However, at the time of these early portrayals of this scene, we can take Jack to be sincere in his belief that his wife was dead. (Comics!)

[7] In fact, the language from *Daredevil* #164 is taken almost verbatim from *Daredevil* #1 (April 1964): "I promised your mother, before she died, that I wouldn't let you grow up to be an educated pug like me! You're going to amount to something, Matt!"

[8] *Daredevil: The Man Without Fear* #1 (October 1993), emphasis added.

[9] *Daredevil* #229 (April 1986).

[10] *Daredevil* #160 (September 1979). In keeping with the theme of promising, Heather asks Matt to promise to put her well-being ahead of being Daredevil, but he refuses to answer, preferring not to make her a promise he knows he cannot keep.

[11] Not all philosophers agree that the deceased are beyond being harmed; for instance, see David Boonin, *Dead Wrong: The Ethics of Posthumous Harm* (Oxford: Oxford University Press, 2019).

[12] Furthermore, although we haven't discussed virtue ethics, it may be the case that keeping a promise to the dead reflects well on us and implies a virtuous character trait such as commitment—another way in which keeping such promises is more about the promisor than the promisee.

[13] *Daredevil* #177 (December 1981).

[14] *Daredevil* #229 (April 1986). Incidentally, around the same time, Matt meets with Spider-Man and details how the Kingpin is ruining his life after learning his secret identity. Although the Kingpin was originally one of Spidey's foes, Matt makes the wallcrawler promise not to intervene, and let Matt deal with Kingpin himself with any "distraction." However, Spidey breaks the promise because he's worried Kingpin might know his secret identity as well, endangering his loved ones (as happened later during the Civil War). Spidey also reflects on Matt, regarding him as being wound a bit too tight, but acknowledging that "he really seemed to care about justice…with a capital J!" (*Amazing Spider-Man*, vol. 1, #277, June 1986).

[15] Making an insincere promise is one of the examples of unethical behavior Kant uses to explain his categorical imperative; see his *Grounding for the Metaphysics of Morals*, trans. James W. Ellington (Indianapolis, IN: Hackett Publishing, 1785/1993), pp. 422, 429-430.

[16] *Daredevil* #177 (December 1981).

[17] Ibid.

[18] For much more on the philosophy of promises, see Allen Habib, "Promises," *The Stanford Encyclopedia of Philosophy* (Winter 2022 Edition), Edward N. Zalta & Uri Nodelman (eds.), at https://plato.stanford.edu/archives/win2022/entries/promises/.

[19] Elsewhere, I have discussed judgment in the context of other superheroes: *The Virtues of Captain America: Modern-Day Lessons on Character from a World War II Superhero* (Hoboken, NJ: John Wiley & Sons, 2014), chapter 5; *Batman and Ethics* (Hoboken, NJ: John Wiley & Sons, 2019), chapter 3; and "Moral Judgment: The Power That Makes Superman Human," in Mark D. White (ed.), *Superman and Philosophy: What Would the Man of Steel Do?* (Hoboken, NJ: John Wiley & Sons, 2013), pp. 5-15.

[20] Immanuel Kant, "On a Supposed Right to Lie Because of Philanthropic Concerns," originally published in 1799, included in the Hackett edition of *Grounding for the Metaphysics of Morals*, pp. 63–67.

[21] Mind you, many philosophers have attempted to untie Kant's Gordian knot: See, for instance, Christine Korsgaard, "The Right to Lie: Kant on Dealing with Evil," in her book *Creating the Kingdom of Ends* (Cambridge: Cambridge University Press, 1996), chapter 5.

[22] *Daredevil* #170 (May 1981).

[23] *Daredevil* #171 (June 1981).

[24] *Daredevil* #172 (July 1981).

Chapter 3: Moral Conflict and Responsibility

Tragic Dilemmas and Threshold Deontology

The problem facing Daredevil with which we ended the last chapter—the choice between letting Bullseye die or letting him live to kill again—serves as a particularly insidious example of the cognitive dissonance that results from a seemingly irreconcilable moral problem. For Matt, there was simply no good choice to be made, because neither option was acceptable. Philosophers refer to this type of choice situation, from which one "cannot escape with clean hands," as a *tragic dilemma*, and it is the superhero's stock in trade, especially one as prone to internal moral conflict as Daredevil.

As it happens, Matt faces another tragic dilemma immediately after he defeats Bullseye, courtesy of the Kingpin himself, who is only starting to play a major role in Daredevil's story. After encouraging Matt to kill Bullseye—and displaying no surprise at all when he chooses not to—Fisk offers him a briefcase full of evidence implicating a number of organized crime figures. If Matt gives the evidence to the authorities, Fisk explains, the resulting cases will eliminate his competition. Finally, just in case Matt doesn't get it, Fisk lays out the problem for him:

> I know what you're thinking, Daredevil. You're planning some desperate, futile attack—you seek to bring me in, as well. You are a very passionate man. But it is not your passion that I now address. It is your intellect.

Consider your position. You have Bullseye—I'll throw him in as a courtesy—and I shall be left with a shattered organization to rebuild. For a time, your side will be that much stronger.[1]

He finishes by appealing to Matt's utilitarian side, all too aware how much it hurts his deontological side: "Consider the greater good to society…and you shall see that you really have no choice, after all." Matt simply says, "You win, Kingpin. This time," and walks away with the briefcase and Bullseye.[2] It will not be long before the Kingpin will become as frequent a source of moral conflict for Matt Murdock as Bullseye has been.

There are several different ways we can look at this situation. One is to regard the Kingpin as using Matt's rule utilitarianism against him, forcing him to violate his rule against letting wrongdoers go free by making the consequences of adhering to that rule too high. Another way is through the lens of *threshold deontology*, introduced by legal philosopher Michael S. Moore in the context of using torture to extract information from terror suspects.[3] According to threshold deontology, we should follow important moral principles until the cost of doing so becomes too high (or reaches a predetermined threshold). In Moore's example, as the potential danger to innocent lives from terrorism increases, at some point even a society that regards torture as wrong in an absolute sense will have to consider if the cost of maintaining their principles is too costly.[4] In other words, how many people should have to die because society—or the select few in charge of it—has decided it will not cross this line?

Even though Matt's decision in the situation above does not have the stakes of Moore's torture example, the Kingpin nonetheless presents his case in terms of the "greater good to society" that would be lost if Matt turned down the evidence and took Fisk in instead (assuming he could, and also that any charges would stick). We can also apply threshold deontology to Matt's repeated choice whether or not to kill Bullseye, which he refuses to do even though this practically guarantees that more innocent lives will be lost in the future. How many people does Bullseye potentially have to kill to make Daredevil seriously consider ending his life? If there is such a number, he presumably hasn't reached it yet—or perhaps Matt does not consider there to be any threshold that would trigger this extreme measure.[5]

Detective Manolis highlights an interesting aspect of threshold deontology when he tells Daredevil that he would be responsible for the innocents killed by Bullseye if he lets him live. In many ways, threshold deontology can be considered more reasonable and practical than the strict variety, given the way it balances both consequences and principles. However, it also implies responsibility for those consequences on the part of those who are in a position to prevent them, even if they played no role in causing them. Matt should not feel responsible for the consequences of doing the right thing; ideally, he would take solace in the fact that he did not take a life, and hold Bullseye responsible for his own actions (assuming he can be judged mentally competent). This is a commonsense ethical position outside of any version of deontology: It is the wrongdoer who chooses to do wrong, and a hero's failure to prevent this wrongdoing does not shift the responsibility for it to them.

But even if Matt understands in his mind that Manolis is wrong about the responsibility for Bullseye's future actions, in his heart he agrees that he does bear at least some of this responsibility. As we know, Matt is not a strict deontologist; he cares too much for consequences for that. It doesn't matter whether we call him a rule utilitarian or a threshold deontologist, because both positions blend deontology and utilitarianism in an unstable combination. Although both schools of ethics agree on the general points about morality, in tragic dilemmas and other cases of moral conflict they inevitably pull in opposite directions, which helps explain why Matt has an outsized sense of responsibility for the consequences of his actions even after he does the right thing according to his principles.

The decision whether to kill his most homicidal foe may be the most dramatic example of moral conflict in Matt Murdock's life, but it is far from the only one. For example, he regularly faces another choice every superhero deals with: whether to pursue an escaping villain or save an innocent person the villain has injured (often as a ploy to escape). As tragic dilemmas go, this choice is easy to make: The hero always stays to save the life in danger at the moment, even though the villain escapes to hurt or kill others. Lucky for us, Matt often reflects on this choice, such as when a man attacked two people in an alley, only to be scared off by a neighbor. By the time Daredevil arrives, he thinks to himself, "I could follow him—I'd love to—but my first duty is to his victims," whom he keeps alive until an ambulance arrives.[6]

Of course, some cases are more complicated, especially when Daredevil is involved. In one story, he and his new friend Wolverine are chasing Bullseye when another killer named Tarkington Brown aims a gun at Daredevil. Wolverine uses his claws to slash Brown's arm as he pulls the trigger, saving Matt's life but nearly killing Brown in the process. Matt barely has a chance to criticize Wolverine's actions before the mutant leaves to chase other criminals, leaving Matt holding the dying Brown while Bullseye gets away.[7]

The situation seems no different from the last one, watching Bullseye escape while Matt tends to an injured victim—except that in this case, Matt knows that Brown is suffering from a fatal illness from which he would likely die soon, even if Matt saves him from dying now from the wounds inflicted by Wolverine. This unique detail would seem to lessen the costs of going after Bullseye, which Matt fully realizes. "Besides," he thinks to himself about Brown, "he's no better than Bullseye...he's a madman...a murderer...he tried to murder me...if Wolverine hadn't stopped him...yes, Wolverine..."[8] It seems that thinking about a hero who kills casually reminds Matt that he's different, and he ends up saving Brown, despite all his valid arguments to the contrary. Generally speaking, he did it for the same reason that he saved Bullseye on the subway tracks earlier: He's not responsible for what these murderers do, nor is he responsible for passing final judgment on life and death, but he is responsible for what he does, and what he does is save lives.

Danger to Loved Ones, Part 1: Karen Page

Another common type of moral conflict that arises for most superheroes is when their costumed antics pose a threat to people they care about. Although his best friend and partner Foggy Nelson is a persistent focus of this, Matt is also concerned about his love interest at any given time: Karen Page in the beginning and Heather Glenn later. (He dates the Black Widow for a while too, but he knows as well as we do that she can take care of herself—as can Elektra Natchios, who presents her own unique moral conflicts, as we've seen.)

Karen found herself in danger almost from the very beginning of her time working for the law firm of Nelson and Murdock, but Matt first references it when she is abducted specifically to lure Daredevil into a trap. After Matt returns to the office and unties his bound and gagged partner, Foggy tells him what happened, prompting Matt to think to himself, "Karen...in danger again! Because of me! It's the one thing— the only thing I've ever feared!"[9] Coincidentally, Matt was planning earlier that day to propose to Karen, but after rescuing her from her abductors, he realizes the folly of this:

Her very life was threatened—because someone merely suspected she knew me! If she became my wife—she'd never be safe for a minute! I love her...too much...to take such a risk! Even if it means...my own life will be...eternally empty![10]

Aside from the element of danger to others, we see also the beginning of an ongoing struggle for Matt Murdock between his duty to use his abilities to help others and his natural human desire for love and companionship. Perhaps more than most superheroes, Matt pursues a number of romantic relationships, some of them reaching the point of marriage (or nearly so), despite his concern with the danger he puts his significant others in. (It also points to the danger of loneliness that accompanies his chosen lifestyle, to which we'll return later.)

Matt also realizes this conflict when it comes to revealing his double life to those close to him, which would make his life tremendously simpler but at equally tremendous cost. After enjoying a workout in his home gym, Matt turns to ruminating about the fact that he cannot confide in anyone about his double life:

I don't dare take Foggy or Karen into my confidence! Sharing my secret could be too deadly dangerous for either of them! I'm not about to jeopardize the safety of my loyal law partner—or the girl I love, just to have an audience![11]

In the same story, after being forced to lie to Karen about his mysterious lengthy absences, he reflects again—and even more very dramatically—on the threat to her safety posed by knowing his dual identity:

Oh, Karen...Karen! If you only knew how I long to take you in my arms... tell you who I really am! But I must not—I cannot...I dare not!! I love you too much to allow you to share my secret—and its dangers![12]

Soon afterward, Karen leaves Matt and the firm because of his refusal to commit to a relationship with her. He escapes once more to his gym, where he considers a possible way to be with Karen and keep her safe, but finds it unacceptable (for reasons left vague):

> How can I jeopardize the life of the girl I love—by letting her share the deadly secret of Daredevil? Only by giving up my crime-fighting career—forever—could I marry Karen—without endangering her life! But I can't!! I just can't do it![13]

This quote points to yet another ongoing struggle often inspired by romantic frustrations: whether to continue as Daredevil or not, which is also tied into his motivations to be a hero in the first place. (We'll expand on these themes later too.)

Karen returns several issues later, but her timing is horrible: Matt learns of an impending attack on Foggy, who's running for New York City district attorney, at their offices. (The attacker is Stilt-Man, who is exactly what he sounds like: a villain who wears a suit with long mechanical legs.) Matt urgently kicks both Karen and Foggy out of the office, earning him the ire of both, but he can only think, "What can I do? What can I say? Foggy's life is at stake!" Later he elaborates on how hard it was to do this, especially at this time:

> It was the toughest thing I've ever done! Just when Karen had returned—when it looked like there was a chance for us again! I had to blow the whole bit! Now, she's probably more hurt—more disappointed—than ever before! And I can't blame Foggy—for thinking I've turned my back on him—when he needs me the most! I can just imagine what they're all saying about me now! But I had to do it! I had to![14]

Next, he thinks to himself, "I had to make certain I'd be here alone—when Stilt-Man attacked! He's too powerful—too deadly—for me to jeopardize the lives of those I love the most!" Despite his tremendous overestimation of the danger posed by Stilt-Man, Matt leaves us with little doubt about how seriously he takes the threat posed by villains to his loved ones due to his double life.

Nonetheless, Daredevil does reveal his secret life to Karen not long after this, but by then things had become more complicated, considering that Matt Murdock faked his own death to foil the plans of a criminal who had learned his secret.[15] It's understandable, then, that the news comes as a shock to Karen in more ways than one—and that doesn't even take into account that Matt chooses to do this at the funeral of Karen's father. Wait, there's more: Matt proposes to Karen, who expresses her own concern about his safety. This is when Matt explains why Daredevil no longer poses a threat to either of them: "Tomorrow, Daredevil has agreed to appear in the United Fund parade....and, I've decided to make it...his farewell performance!"[16] Unfortunately, during the parade, the Stunt-Master tries to kill him, and after thwarting his attempt, Daredevil decides to stick around—and Karen decides to leave (again).

Although she returns to Matt's life (again) soon thereafter, Karen continues to struggle with his double life, eventually leaving for good, not to return to his life until she hits rock bottom and is forced to sell the secret he shared with her. As for Matt, this episode represents not only his ongoing heartache and discomfort with letting people in on his secret, but also a repeated back-and-forth between his two identities in which he cycles between being Matt and being Daredevil, and occasionally quits one or the other. (More on both of these aspects near the end of the book.)

Danger to Loved Ones, Part 2: Heather Glenn

We turn now to Heather Glenn, Matt's first serious love interest following Karen and Natasha. Besides protecting her from his foes, Matt also has to deal with her father Maxwell, who is often mixed up with criminal enterprises (while at the same time helping to fund Matt's legal aid clinic). Matt suspects the elder Glenn is lying when he claims to be innocent of any wrongdoing, so he prepares to investigate while anticipating the personal costs. "I knew that someday I'd have to do something which would hurt the one I love," he thinks to himself. "I fought against it, desperately wishing that day would never come... but here it is, slapping me in the face." In his head, he lays out the heart of the moral conflict: "If Maxwell Glenn is guilty, it will crush Heather. But if I do nothing to learn the truth, it will destroy me—not to mention corrupting the meaning of justice."[17]

He expresses his feelings about his own role and responsibility in the situation, not only regretting that Heather might be hurt, but also resenting that it falls on him to do it:

> Yet, whenever I think of Heather and that nutty, contagious joy she always brings to everything, I wonder why I should be the one who has to hurt her. Who appointed me guardian of all that's right and wrong?

He concludes his "soliloquy" (as he calls it later) by remembering his purpose, regardless of how he feels about it:

> But, when I'm finally disgusted with being Daredevil, I remember why I am what I am. I remember all the people I've helped. And I know why I do what must be done. Sometimes you've got to do a job you hate because there's no other choice. And maybe, just maybe, that's what sometimes makes me the loneliest man in the world.

We'll revisit this passage later when we discuss Matt's motivation to be Daredevil, as well as his persistent feelings of loneliness. For now, it's enough to see that he eventually acknowledges his responsibility to do the right thing even when it's inconvenient for those he loves—and also for him.[18]

Eventually, Matt learns that the Purple Man was manipulating Maxwell Glenn the entire time, even convincing Glenn himself he's guilty, and realizes he could resolve the situation quickly and easily by speaking to Heather and Foggy as Daredevil. But he considers that to be too risky, admitting to himself that "I'm always worried that Foggy will finally see through my disguise…or Heather would recognize me—and maybe leave me like Karen Page did!"[19] We see here that physical danger is not the only problem that his secret identity creates for Matt and the people in his life—and this time the complications it creates are enough to make him decide to tell both Heather and Foggy the truth, feeling guilty especially for not telling his best friend earlier: "I should have trusted him long ago."[20] But he doesn't get a chance to follow through with this, because while he is on the phone, learning that Maxwell has taken his own life in prison, Heather bursts in and sees him in costume but unmasked. At this point, she realizes that the man she loves is also the man who sent his father to jail (and, effectively, to his death).[21]

Heather is understandably devastated and enraged, but Matt is even harder on himself, feeling responsible for Maxwell's suicide, given how involved he was in the case and how compromised he was because of his dual identity: "I—only wanted to help, but instead I failed everyone who ever meant anything at all to me!…And I've no one to blame…except myself." While taking out his own anger on his apartment later, he rants about the costs of his secret life and swears once again to give up being Daredevil forever. But after thwarting bus hijackers later as Daredevil, he realizes that despite the mistakes he's made in both of his identities, he's also done a lot of good in both. "That may not justify the mistakes," he thinks, "but it does make them a lot easier to live with."[22]

Years later, after they broke up, Matt has a final interaction with Heather Glenn that further illustrates the conflicts brought about by his outsized sense of responsibility. After she calls him for help, he rushes to her apartment, only to learn she is drunk and faked the distress call to lure him over. Unfortunately, Matt had been in such a hurry to get to her that he declined to intervene in a violent domestic dispute along the way, and after leaving her, he discovers that the dispute turned fatal. (Somewhere, Peter Parker simply nods.) When Heather calls him again, Matt tells her, "Because of you, I didn't do something I should have done...and a woman died!" Heather replies, "I'll die too," but her voice cuts out when Matt yanks the phone out of the wall in frustration.[23]

Soon afterward, Matt learns that Heather took her own life, and he feels responsible for that too, thinking it was "my fault we broke up...My fault she got so messed up...If I'd tried to understand her," and so on. After Foggy reads her suicide note to Matt and asks him how he feels, he answers, "Guilty. Guilty as sin." When Foggy explains that Matt just didn't love Heather as much as she loved him—which Matt had thought to himself earlier—and that he shouldn't feel guilty about that, Matt says, "Maybe not. But still, she's dead...and in some way, I'm responsible. It's my fault."[24] After Heather's funeral, when Ben Urich asks him, incredulously, if he feels responsible for Heather's death, Matt rehearses all the things he could have done and all the things he couldn't, and concludes, "I don't know. I'll never know."[25]

When he characterizes himself as a failure and assumes responsibility for tragic events without having played a direct role in them, Matt shows us how much responsibility he takes on his shoulders, regardless of whether he bears any moral responsibility as a philosopher would think of it. He even feels responsible for everything that happens to Elektra, despite being in her life a relatively short time, asking himself at one point, "What made me hurt you so? Did I fail you in some way—or were you possessed by some awful inner demon that made you kill, and kill..."[26] (The safe bet is on the latter.) Like many heroes, Matt tries to do the right thing, but even when he does—and does more than anyone could possibly be expected to do—he still feels responsible for the harm he couldn't prevent, and especially the harm that occurred as a byproduct of his heroic endeavors. This attitude may be emblematic of heroes in general, whether "super" or not, but for someone as internally conflicted as Matt Murdock, it has a definite negative impact on his mental state, as we will see later in this book.[27]

[1] *Daredevil* #172 (July 1981).

[2] Ibid.

[3] See Michael S. Moore, "Torture and the Balance of Evils," *Israel Law Review* 23(1989): 280–344. For a critique of threshold deontology, see Larry Alexander, "Deontology at the Threshold," *University of San Diego Law Review* 37(2000): 893–912.

[4] This assumes, against copious evidence, that torture is effective at generating truthful information; see the extended discussion of torture below.

[5] This tragic dilemma is faced by many superheroes who fight particularly homicidal foes. A prime example is Batman and his arch enemy the Joker, which I discussed briefly in "Why Doesn't Batman Kill the Joker?" in Mark D. White and Robert Arp (eds), *Batman and Philosophy: The Dark Knight of the Soul* (Hoboken, NJ: John Wiley & Sons, 2008), pp. 5–16, and in more detail in chapter 6 of *Batman and Ethics* (Hoboken, NJ: John Wiley & Sons, 2019).

[6] *Daredevil* #173 (August 1981).

[7] *Daredevil* #196 (July 1983).

[8] Ibid.

[9] *Daredevil* #29 (June 1967).

[10] Ibid.

[11] *Daredevil Annual* #1 (1967), "Electro, and the Emissaries of Evil!" (Yes, it does say "deadly dangerous." I couldn't make that up.)

[12] Ibid.

[13] *Daredevil* #43 (August 1968), an issue that also features a fight against Captain America, discussed here: https://thevirtuesofcaptainamerica.com/2018/11/05/captain-america-103-104-and-daredevil-43-july-august-1968/.

[14] *Daredevil* #48 (January 1969).

[15] Matt faked his death in *Daredevil* #54 (July 1969) and revealed his dual identity to Karen in *Daredevil* #57 (October 1969).

[16] *Daredevil* #58 (November 1969).

[17] *Daredevil* #143 (March 1977). (The next two blockquotes also come from this issue.)

[18] We also see a hint of his self-obsession, which will become more apparent as his life beginning to fall apart even more, as we discuss later.

[19] *Daredevil* #150 (January 1978).

[20] Ibid. In fact, Foggy will not learn Matt's dual identity until December 1995's *Daredevil* #347 (and only after Matt fakes his death *again* to protect his secret).

[21] *Daredevil* #151 (March 1978). (The rest of the events and quotes in this paragraph come from this issue.)

[22] His retirement as Daredevil may not have lasted the issue, but it did seem to help him forget his decision to tell Foggy his secret!

[23] *Daredevil* #220 (July 1985).

[24] Ibid.

[25] *Daredevil* #221 (August 1985).

[26] *Daredevil* #182 (May 1982).

[27] For other examples of superheroes assuming more responsibility than they deserve, see my work in *The Virtues of Captain America: Modern-Day Lessons on Character from a World War II Superhero* (Hoboken, NJ: John Wiley & Sons, 2014), pp. 58–63; "The Otherworldly Burden of Being *the* Sorcerer Supreme" in Mark D. White (ed.), *Doctor Strange and Philosophy: The Other Book of Forbidden Knowledge* (Hoboken, NJ: John Wiley & Sons, 2018), pp. 177–190; and "Panther Virtue: The Many Roles of T'Challa," in Edwardo Pérez and Timothy E. Brown (eds), *Black Panther and Philosophy: What Can Wakanda Offer the World?* (Hoboken, NJ: John Wiley & Sons, 2022), pp. 53–60.

Chapter 4: Law-Breaking, Violence, and Torture

Breaking the Law

Although we can see the importance and influence of rules throughout Matt Murdock's life, including the moral code he holds himself to, no rules are more prominent than laws themselves, in both his public life as an attorney and his private life as a masked crimefighter. Most superheroes run afoul of the law from time to time, of course, and some do so more than others, but few embody this conflict as intensely as Daredevil. (Even Jennifer Walters, the Sensational She-Hulk, seems to treat her day job as an attorney more like an ordinary job than a calling.)

As problematic as this contrast may seem to the reader, Matt doesn't seem to take it as seriously as one may expect, at least not in his early years. In one notable case during which he comments on it, in the process of fighting the Death-Stalker, he finds scientific plans to make people into monsters immune to pollutants, which he feared would be used to create super-soldiers resistant to chemical warfare, thereby increasing the threat of war. He felt this information was too dangerous to fall into the hands of any government, "including my own," and proceeds to destroy them.[1]

The papers weren't just a danger to the world, though, but also state's evidence in a case close to Matt's heart—or, at least, close to the heart of someone close to his heart. As he leaves the scene, Matt reflects on the fact "that I deliberately destroyed the government's evidence against Foggy Nelson's sister Candace. And yet...I had to. No nation, not even this one could be trusted with the secret those papers contained."[2] He continues to ponder the ethical aspects of what he did:

> Morally, I owed it to the whole world to do what I did. But legally—I've committed a crime, no matter how justified. And more than that, I've acted against the whole system of laws...which Matt Murdock swore to uphold.[3]

Very rarely does Matt articulate this as clearly as he does here, which may suggest that he has internalized the conflict so completely that he doesn't even think about it in the context of "everyday" law-breaking. After all, this is hardly the first time Daredevil has "committed a crime," but this particular crime would be obstruction of justice, one that, given his day job, may hit him a bit harder than routine violence and breaking-and-entering. (And looking ahead a bit, we'll see that the perpetual see-sawing between whether he should or should not continue as Daredevil never seems to hinge on the illegality of many of his nighttime activities.)

Another reason Matt may not be as conflicted about breaking the law as we might expect lies in a passage quoted earlier from the same story, in which he laments that "even the one thing I always had real faith in— the law—even the law seems to work backwards these days."[4] Matt realizes all too well the distinction between law and justice, not despite his day job but because of it. As an attorney, he works within the law to secure justice for his clients, but the outcome isn't always consistent with justice in an ideal sense because of the imperfect nature of law (as discussed earlier). As Daredevil, though, he can work outside the law, allowing him an alternative means to pursue the same end, justice (but for a much wider group of people than merely his clients).[5]

Violence

Of the various ways Daredevil runs afoul of the law, perhaps the most significant, frequent, and morally disturbing is his penchant for using violence, often in extreme ways, when fighting crime. Of course, some violence is expected and unavoidable when "engaging" with criminals, especially those with enhanced powers or dangerous equipment. It would be very difficult to do battle with the incredibly strong Mr. Hyde, the sharpshooter Bullseye, or the…very tall Stilt-Man without resorting at some point to hand-to-hand combat. Even though this is what he promised his father he wouldn't do, which we addressed in chapter 2, this type of fighting is not our main concern here. Instead, we are going to focus on his use of violence against individuals who have not necessarily done anything wrong, but are merely useful to Daredevil for some reason—usually because they might have information he needs to find those who have done wrong (or threaten to do so).

Although Daredevil has a reputation today as a violent hero, this was not always the case. In his earlier days, it was more common to see him criticizing others for being too violent, as he did with Black Widow early in their partnership when they stop a bunch of men from mugging an elderly woman outside a welfare office. The men pose no challenge for the heroes, so Daredevil toys and banters with them while incapacitating them, only to see the Widow repeatedly beating one of the men, calling him "filth" and saying, "He deserves to be hurt… to feel pain." After Daredevil stops her and the police arrive, Natasha disappears, leaving Matt to reflect on her behavior, sensing "more than anger—real hatred to me—frighteningly intense." Later they reconnect and she defends herself, citing the men's brutality, to which Matt replies, "And you sink to their level. How admirable." Matt wonders later, "Can I love a woman who can be so—vicious?"[6]

It takes several years after this for Matt to embrace a similar degree of violence himself, and ironically it is Natasha who points it out. Seeking revenge on Daredevil, Bullseye subdues and abducts the Black Widow to draw him out (although they are no longer partners professionally or romantically). Even though she manages to free herself before Matt shows up, he still unleashes his rage on Bullseye, prompting Natasha to think to herself, "I don't think I've ever seen Matt so grim…so cold-blooded. He's changed, he's not the same man I used to know."[7]

There is no single explanation for this change in Matt Murdock's attitude toward violence. More likely it is the cumulative effect of much of what we have discussed so far in this book, namely the many overlapping frustrations of leading his complicated life. He is a man who constantly tries to do the right thing and follow the rules (and the law), only to see things become steadily worse. He holds himself responsible for protecting the innocent residents of Hell's Kitchen while also keeping his loved ones safe, but can't seem to do both, so he writes himself off as a failure—which, given the impossibly high standard he's set for himself, is practically guaranteed to continue. And let's not forget his "normal" life as Matt Murdock, in which his employment status and his romantic life seem to be constantly in flux, largely due to his superheroic activities. Against all this turmoil, the one thing Matt can rely on is his fists—okay, make that two things—so it makes sense that he would embrace this as a way to forget his troubles. And it doesn't help that he repeatedly has to deal with a uniquely psychopathic, homicidal, and formidable enemy such as Bullseye, making his turn to extreme violence all but inevitable.[8]

It is not only when dealing with homicidal psychopaths like Bullseye that Daredevil is seen to become more violent. At some point between calling out the Black Widow for her extreme violence and doing the

49

same in front of her, Matt began beating up random thugs for information. As I previewed above, it is one thing for him to respond to violence with violence, or to use violence to bring in a confirmed and dangerous lawbreaker (especially a violent one), but it is another thing entirely for him to use violence against someone who, as far as he knows, has done nothing wrong, just because they may able to provide useful information.

Nonetheless, he does exactly that, beginning with Archer Emmet, a drug dealer who claims to have gone straight (although our hero has his doubts). Matt doesn't want to take Archer to the authorities, or even question him about any recent criminal activity—instead, he combines his insinuations about Archer's current illicit dealings with physical threats. After asserting that Archer was still dirty even though the police have yet to prove it, Matt pushes him down onto a table and says, "Fortunately for me, I'm not a cop. So I can break your face into a jigsaw puzzle if I want to—unless you give me the info I need." Archer says he doesn't know anything, which Matt believes (without making reference to listening to his heartbeat), but before he leaves he makes a final threat: "If I find out you're lying to me, better send out for a couple thousand band-aids for your face. You'll need them."[9]

As he leaves, Matt thinks to himself, "I hate putting on the tough-guy act, but those guys need it, or they wouldn't raise a finger to save their grandmothers!"[10] To be fair, the violence here was mostly an act, consisting of one shove to get Emmet's attention, followed by threats that he may not have acted upon. But it is notable that at this point, Matt regrets even this much, because it doesn't take long for him to actually beat up some organized crime figures he hopes will lead him to Zebediah Killgrave, the Purple Man, who used his mind control powers to make Maxwell Glenn commit the crimes for which he went to jail.[11] Again, these men may have committed crimes and were certainly ready to use

50

their weapons, making them legitimate threats and justifying a response. The important point is that Matt was not there to pursue any existing charges against them—he had his own reasons for roughing them up that had nothing to do with any criminal wrongdoing on their part.

Soon this behavior becomes somewhat of a trademark for the Man without Fear, especially when it happens in a fine drinking establishment. In fact, the next time we see him beat up thugs for information, it is in the famous Josie's Bar, making its first appearance (albeit on Manhattan's Lower East Side, not Hell's Kitchen), when he is looking for Bullseye after he abducted the Black Widow. For five pages, first in disguise as just another tough guy and then later as Daredevil, Matt thrashes nearly everyone in the bar (except Josie), trying to get information leading him to Bullseye. The second to last panel shows Matt holding up two-bit lowlife Turk Barrett by his shirt collar, telling him to give Bullseye a message for him, while numerous customers of Josie's lie unconscious at his feet.[12]

Maybe we're being a bit harsh here. After all, if these people have information that will help Daredevil catch a dangerous criminal like Bullseye, what's the problem with roughing them up a little to get it? The problem has less to do with the violence itself and more with what it implies about how he's treating the people he uses it on. You'll remember Immanuel Kant, the deontologist we discussed in the first two chapters, who devised a test called the categorical imperative for testing plans of action for their moral status. The categorical imperative comes in several forms—all logically equivalent, according to Kant—one of which is the Formula of Respect: "Act in such a way that you treat humanity, whether in your own person or in the person of another, always at the same time as an end and never simply as a means."[13] In other words, we should not use people for our own purposes without also treating them as individuals worthy of respect and consideration. Among other things, this clearly forbids actions such as lying, which treats the person lied to like a tool to get whatever the liar wants, and coercion, in which a person is forced to do something against their control to benefit someone else.

Now, if Daredevil were beating up Turk in order to capture him and turn him in for a crime he was legitimately suspected of committing, that would be one thing. In that case, Matt would be holding Turk responsible for his (suspected) actions, which treats him with respect insofar as it recognized Turk as an autonomous person. But it's another thing entirely to beat him up simply as a way to get information about someone else. As we said before, it doesn't matter that Turk very well may have done something to arouse the attention of the law, because that isn't why Matt roughs him up. Matt literally uses Turk as a means to his own ends, finding Bullseye and rescuing Natasha, which denies Turk the basic respect that every person is owed by virtue of being a person who can choose their actions and take responsibility for them.

Torture

After this point, in almost every issue Daredevil beats up random thugs—even some who aren't named Turk—for information that may lead him to someone he actually cares about stopping or catching.[14] But if his violence against people who are merely useful to him isn't bad enough, he crosses another line when he threatens a thug named Mickey for information. He doesn't hurt Mickey, but he does dangle him off a roof and threaten to drop him if he doesn't talk: "It's a long way down, rat. Get smart." Mickey calls his bluff, telling Daredevil, "You got a rep—everybody knows you ain't never killed nobody!" So Matt drops him, then swings down to catch him just before he hits the ground, saying, "That's true... but I can keep this up all night. Can you?" As planned,

Mickey talks (after he empties the contents of his stomach on the ground below).[15]

It is not actual pain that convinces Mickey to talk but the threat of it—and not just any pain, like getting punched in the face, but the terrifying prospect of falling to his death. In other words, what Daredevil does to Mickey easily qualifies as torture, which can be defined as "any act by which severe pain or suffering, whether physical or mental, is intentionally inflicted on a person for such purposes as obtaining from him, or a third person, information or a confession."[16] Although Daredevil's normal beatdowns in Josie's Bar are certainly violent and painful, threatening someone with a multi-story plummet to their death is on another level, and its uniquely psychological nature places it squarely in the category of torture. (The fact that Mickey believes Daredevil won't actually kill him is immaterial: Even people who know that waterboarding only simulates the experience of drowning often experience extreme distress from being subjected to it.)

We shouldn't take the moral case against torture for granted. After all, we may look at it like violence, something that is ordinarily regarded as wrong but becomes a "necessary evil" when confronted with evil itself.[17] Although Matt may engage in battle with over-muscled henchmen or even master combatants like Bullseye, his use of torture is usually applied to the same people he beats up for information—people who have not necessarily done anything wrong and did not threaten Daredevil with violence, but who might have information he needs. This use of violence uses these individuals merely as means, but torture takes it up a notch, not just using people but almost literally using them up. As philosopher Bob Brecher writes, in the worst cases "torture breaks people," robbing them of their ability to make free choices—and without that, "the tortured subject is no longer a person."[18] Add to this the fact that torture rarely results in useful information, because the tortured person

53

will say anything to make it stop, and we see why torture is condemned on both deontological and utilitarian grounds.[19]

Daredevil may not engage in torture as often as other costumed crimefighters who shall remain nameless.[20] But it is concerning nonetheless, especially given Matt's early distaste for violence and his ready acceptance of it, including torture, not long into his career. But even just the violence itself is bad enough—so bad that the Kingpin himself has to point it out to Matt. After Bullseye kills Elektra, Matt refuses to believe she is truly dead; as part of his search, he fights his way through Fisk's security to accuse him of hiding her. After Fisk inquires after his mental health, Matt answers, "I'm fine. Now, do you tell me the truth—before, or after I beat you senseless?" Amused, Fisk says that "the Daredevil I know would never resort to unprovoked violence simply to test a theory. Would he?" Matt leaves, less to prove the Kingpin wrong than out of recognition that he's right: The Daredevil we know would "resort to unprovoked violence" for much less than a theory.[21]

Second Thoughts?

Even if Matt did not fully appreciate what the Kingpin said above, he does find cause to question his violence in a pivotal tale about the behavior he inspires in others.[22] Daredevil actually tells this story to Bullseye, who is lying paralyzed in a hospital bed after their intense battle following Elektra's death at Bullseye's hands. (We'll discuss the battle more below.)

The story starts when Matt Murdock visits a new client, Hank Jurgens, and meets his young son, Chuckie, who is obedient to his domineering father but idolizes Daredevil. Later, Daredevil picks up Chuckie at school and shows him what parkour really is, and Chuckie tells him he's his hero because "when somebody gets in your way, you give it to them—POW." Matt doesn't immediately appreciate what Chuckie meant or why exactly he looks up to him, which backfires when Daredevil catches Hank in a criminal act and punches him out—in front of Chuckie.

Chuckie has a mental break, deciding that if his hero Daredevil hit his father, his father must be bad—not an unreasonable conclusion—but then Chuckie decides that he must be bad as well. Later, when his dad is sentenced, other kids tease him on the playground, and Chuckie pulls out a gun and shoots one of them in the arm. Chuckie disappears into himself, leaving Matt to wonder how responsible he is for what Chuckie did. "What am I giving people, by running around in tights and punching crooks?" he asks Bullseye. "What am I showing them?" He continues:

> Am I showing them that good wins out, that crime does not pay, that the cavalry is always on its way—or am I showing them that any idiot with fists for brains can get his way if he's fast enough and strong enough and mean enough? Am I fighting violence—or teaching it?

Next, Matt tells Bullseye the story about getting into a fight as a boy and his father hitting him for it, the event that taught him the importance of rules in helping good people avoid doing bad things and set him on the path to becoming a lawyer.

But he never really answers his own question about the message he's sending with his violent behavior as Daredevil. The tragic case of Chuckie is too complicated to lay at Matt's feet, likely having more to do with his harsh and controlling father than with anything behavior modeled by Daredevil. Nonetheless, this situation provides Matt with a chance to reconsider his violent methods through the impression it makes on other people…but unfortunately it is a chance that goes un-appreciated, given his continued use of extreme violence for questiona-ble ends.[23]

Perhaps Matt dismisses this concern so quickly because he doesn't really see himself as a public hero who needs to maintain a clean image to serve as a role model (or what philosophers call a *moral exemplar*). Matt doesn't see himself as setting an example for others—this is a role he would more likely embrace in his private life as an attorney (and a blind one at that), dedicating his life to using the legal system to help others who cannot help themselves. As Daredevil, he serves a similar purpose but in the shadows rather than the courtroom. Although he oc-casionally works with superhero teams, he never joins them—despite several invitations, including from the Black Panther to join the Avengers, which Matt declines by saying, "I've always worked alone."[24] Although this is true, one reason for it may be to avoid the spotlight that seems to follow the Avengers and other teams around, casting their members as public figures, which suits traditionally inspirational heroes such as Captain America much more than Daredevil.[25]

[1] *Daredevil* #115 (November 1974).

[2] *Daredevil* #116 (December 1974).

[3] Ibid.

[4] *Daredevil* #115 (November 1974).

[5] In this way he represents a 1980s hero from DC Comics called Vigilante, also known as Adrian Chase, a judge who grew tired of seeing criminals acquitted on technicalities and decided to start enforcing his own brand of justice outside the courtroom. (Costume aside, the Adrian Chase/Vigilante character in the HBO Max series *Peacemaker* has little to do with the comics version.)

[6] *Daredevil* #108 (March 1974). For an example of extraordinary violence before he became Daredevil from the miniseries *The Man Without Fear*, see the beginning of the next chapter.

[7] *Daredevil* #161 (March 1974).

[8] Bullseye first appears in *Daredevil* #131 (March 1976), although he doesn't come into his own until *Daredevil* #159 (July 1979).

[9] *Daredevil* #139 (November 1976). Note that this is over two years before Frank Miller joins the title, in case you thought it was he who introduced this behavior to Daredevil's arsenal.

[10] Ibid. Fun fact: Matt actually beats up quite a few people in the rest of this issue, which was no act.

[11] *Daredevil* #148 (September 1977).

[12] *Daredevil* #160 (September 1979). (By this point, Miller is on board.)

[13] Immanuel Kant, *Grounding for the Metaphysics of Morals*, trans. James W. Ellington (Indianapolis, IN: Hackett Publishing, 1785/1993), p. 429.

[14] Bullseye does this too, which should give Matt pause: For instance, see *Daredevil* #172 (July 1981), when Bullseye roughs up some of Matt's favorite informants himself for information about the Kingpin.

[15] *Daredevil* #168 (January 1981). Earlier in this issue, Daredevil beat up a guy for information who was (ironically) masquerading as a blind man to escape notice. See if you can guess his name! (Hint: It rhymes with *jerk*.)

[16] This comes from the 1984 United Nations Convention against Torture and Other Cruel, Inhuman or Degrading Treatment or Punishment (Part 1, Article 1, at https://www.un.org/documents/ga/res/39/a39r046.htm). On the difficulty of defining torture, see section 1 of Seumas Miller's entry on the subject at *The Stanford Encyclopedia of Philosophy* (Summer 2017 Edition), Edward N. Zalta (ed.), at https://plato.stanford.edu/archives/sum2017/entries/torture/.

[17] We saw another possible justification in the form of threshold deontology in the last chapter, but that argument depends on there being catastrophic costs to not using torture, which would not apply to Daredevil's typical cases (however passionate he feels about them).

[18] Bob Brecher, *Torture and the Ticking Bomb* (Malden, MA: Blackwell, 2007), pp. 75 and 77.

[19] See ibid., pp. 24–31; Yvonne Ridley, *Torture: Does it Work? Interrogation Issues and Effectiveness in the Global War on Terror* (Saffron Walden, UK: Military Studies Press, 2016); and Jeannine Bell, "'Behind This Mortal Bone': The (In)Effectiveness of Torture," *Indiana Law Journal* 83(2008): 339–361.

[20] There's a reason I spent ten pages discussing torture in *Batman and Ethics* (Hoboken, NJ: John Wiley & Sons), pp. 169-179.

[21] *Daredevil* #182 (May 1982). Bullseye killed Elektra in the preceding issue, just thirteen issues after her introduction in *Daredevil* #168 (January 1981).

[22] All details and quotes that follow are from *Daredevil* #191 (February 1983) unless otherwise noted.

[23] He does, however, meet up with Wolverine soon after this, and long before the episode with Tarkington Brown described above, he comments on Wolverine's brutality after he threatens Turk with his deadly claws (*Daredevil* #196, July 1983).

[24] *Daredevil* #99 (May 1973). He quickly adds "more or less" because his crime-fighting partner at the time, the Black Widow, was *right there.*

[25] Cap is definitely *the* moral exemplar in the Marvel Universe, as I detail in *The Virtues of Captain America.*

Chapter 5: Killing

The Early Years, the Black Widow, and the Punisher

Daredevil may engage in frequent acts of violence—often extreme, and sometimes lapsing into torture—but like most other superheroes, he rarely kills. In fact, he makes a point of this on many occasions, holding others to the same rule, refusing to break it even when he knows it would prevent more innocent deaths in the future, and especially when he admits to himself and others that he truly wants to.

Matt was not so careful about this in his earliest years before he "officially" became Daredevil. In the process of pursuing the three men responsible for killing his father, he beats the first one senseless (including kicking him in the ribs repeatedly after he was down) and chases the second one into a subway and watched him have a heart attack. While closing in on the third, partying with a number of business associates and scantily clad women in an apartment, Matt accidentally knocks one of the women out of a window. He shouts, "No...no...no!" while the exposition very emotively reads: "She's dead! He killed her! He didn't even know her and he killed her!"[1]

Although this inadvertent death affects Matt deeply—likely because she was an innocent who was not involved in any way in his father's murder—before long he directly causes two deaths, although there were extenuating circumstances in both cases. When a young girl he had befriended is kidnapped, Matt adopts a disguise for the first time—specifically, the black bandanna covering his eyes, which was made famous in the first season of the Netflix series. He clears a bloody path through the

New York City underworld, delivering vicious beatings to countless thugs in search of his friend, until he tackles two men off a pier into the water. One quickly drowns, but the other pulls a knife, which Matt stabs him with, alongside narration that reads: "A knife—no choice—give it back to him."[2] This is a clear example of self-defense, and no one would blame him for this, especially because he was underwater and did not have time to explore other options.

When he finds a gunman with the girl, holding a gun to her head, Matt pleads with him, over and over, "Let her go. I don't want to kill you." The man shoots repeatedly at Matt, most of the bullets missing him or being repelled with his billy club. One does find its way into Matt's arm, but still he says, "Please. I'm begging you. I don't want to kill you. Let her go." Finally, Matt swats the man's last bullet back between his eyes, killing him instantly and saving the girl.[3] Once again, Matt was unarmed, facing off against a man with a deadly weapon, and trying to save a young girl's life without losing his own. It is not hard to grant him the benefit of the doubt that he waited until the last minute before he ended the gunman's life. Nonetheless, later in Matt's career as Daredevil, we would expect him to find another way to defeat his enemies without resorting to deadly force.[4]

Once he is established in his career as Daredevil, Matt avoids killing and expresses concern when other people seem to consider it. Appropriately it is the Black Widow who triggers an early alarm, as she did earlier with her penchant for violence: When she threatens to kill the villain Angar unless he stops producing dangerous mass hallucinations, Matt reflects later on his concerns about her: "She wasn't bluffing Angar! She really would have killed him. And somehow that scares me—a lot."[5]

But it is another person in black whom Matt confronts directly over his habit of killing: Frank Castle, the Punisher, who fights a lethal one-man war across criminals. When he first encounters Castle in action,

Matt gets a clear mental picture of him: "He's big. His heartbeat is strong, his movements confident, fluid. He's in excellent shape. And he's a killer!"[6] When he asks Ben Urich later what he knows about Castle, the reporter tells him, "Haven't got much on him. He's a vigilante, like you...," but Matt is quick to correct him: "Not like me. He kills."[7] To drive the distinction home even further, when Matt finds Castle violently questioning a thug on a rooftop, he steps in. When Frank tells him, "You have your methods, Daredevil. And, so do I," Matt replies (without a hint of irony), "Mine don't include senseless brutality. Or wanton murder."[8]

Daredevil and the Punisher have encountered each other many times since, and almost every time they discuss their different views on killing. The die was cast in their first meeting on the rooftop, which continues when Frank suggests he and Matt work together to eliminate crime once and for all. Matt's answer is unequivocal: "Whether you kill innocents or criminals, it's murder—and that makes us enemies, Punisher."[9] Setting aside the question of whether either man is qualified to declare anybody guilty—which Matt Murdock, attorney-at-law, knows the answer to—the importance of this reply is that Matt regards any killing as unjustified homicide, regardless of whether a person is a criminal or how many deaths they may be responsible for. Even if he directly witnesses a crime, including a murder, Matt realizes it is the job of the criminal justice system to hold the perpetrator accountable, not his—and certainly not Frank Castle's.

From this point on, Matt takes it as his mission to save the lives that the Punisher threatens. During the same episode, Frank beats a man until his heart stops and Matt tries to revive him, saying, "His heart...got to massage...try to start it beating again...undo the damage the Punisher did." He thinks to himself, "The Punisher won't take a life," then says aloud: "Not. Another. One!"[10] When next they meet—again on a roof-top—Matt tries to stop Frank from shooting someone else, and when Frank shoots toward Matt instead, he says,

> You don't scare me, Punisher. I've done my homework—and I know you're missing me on purpose. You've killed dozens of criminals in your time. But you've never harmed an innocent. You won't kill me.[11]

Again, ignoring whether they can accurately identify innocents and criminals—an issue in any serious argument over vigilantism—Matt hits on perhaps the only part of Frank's moral code that keeps him from being cast as a full-blown villain and makes Matt feel comfortable working with him when necessary. (Interestingly, Matt finally stops Frank by picking up a gun that belonged to one of Frank's earlier victims and shooting Frank in the shoulder with it, one of the few times Matt is seen using a gun.)[12]

Bullseye and the Trolley Problem

Daredevil's arguments with the Punisher may have cemented his position against killing, but Bullseye gets the honor of testing his resolve—a test that Matt has not always passed. He starts out well: When a doctor tells Daredevil and Manolis that Bullseye has a brain tumor, which will kill him if not removed soon, Manolis says sarcastically, "That'd be a real loss to society," to which Matt replies, "Any death is a loss, Manolis." When the detective scoffs and says, "Nuts. Killing is like breathing to that slime. He doesn't deserve to live," Matt explains: "That's not for either of us to decide. We have to save him."[13] Here, Matt asserts, as he did to Frank, that all life is valuable, regardless of guilt or history, and also that the job of delivering judgment belongs to someone else, which may refer to either the courts (in this world) or God (in the next).

Matt shows his devotion to saving lives when Bullseye takes a hostage and makes an offer: "Drop your billy club—stand stock still and let me kill you—or I'll skewer this little creep! It's your choice, devil! His life—or yours!" Matt calls his bluff and lets Bullseye fling a blade into his shoulder. As he did with Frank, Matt takes preventing further deaths personally, telling Bullseye when he finds him again, "I won't let you kill again, Bullseye! I won't!" Nonetheless, as we saw earlier, when he faces the choice of whether to let Bullseye die under the wheels of a subway car or pull him to safety, Matt finds he has no choice but to save him, even though letting him die would prevent countless more deaths in the future.

As with many similar choices that often confront superheroes, here Matt faces his own version of the famous "trolley problem." In the classic version of this philosophical thought experiment, a trolley car carrying five passengers is heading down a track towards a broken bridge, which would plunge the trolley into the ravine below, killing all five passengers. However, the trolley car can be diverted onto another track where a single person is standing; if the trolley is switched to the second track, the one person on it would be killed but the five passengers would be saved.[14]

The trolley problem poses the surprisingly complex question: If you had the chance to pull the switch and divert the car, would you? If the only relevant factor to you is the number of people killed, then you would probably say yes, because saving five people justifies the loss of one. But even if you acknowledge this, there is still a catch: If you pull the switch then you have acted to kill the one person on the other track, whereas if you do not pull the switch, you have not inserted yourself into the situation at all, as if you were never there. Many philosophers draw a sharp distinction between actively causing a death and passively allowing a death to occur, so even though this second option results in more deaths, it absolves you from direct involvement in any of them. After all, we can assume that other people were more directly responsible for the trolley being in that situation, and perhaps, like Dante in the movie *Clerks*, you weren't even supposed to be there today. On the other hand, by walking away you would be giving up the opportunity to save the five passengers—does this make you responsible for their deaths in any way? The trolley problem raises many questions, which is what makes it a fascinating situation to ponder.[15]

Superheroes face a version of this problem whenever a dastardly villain gives them a choice between saving their one true love or a bus full of schoolchildren—but because they're superheroes, they can usually find a way to save everybody, thwarting the tragic dilemma they were presented with. They sometimes also face the tougher decision Matt is confronted with here, in which they have the option of killing their most murderous foes (or refusing to save their lives), preventing the many deaths they would likely cause if they were allowed to live. And like most of his costumed colleagues, Matt chooses to save the lives in front of him, despite any murderous inclinations they may have, even at the expense of lives in the future. In this way, he goes a step further in his opposition to Frank's methods: Not only does he refuse to kill Bullseye, but he actually saves him when he could simply turn away and let the subway car take his life.

Why does Matt make this choice? We don't know for sure, but one good guess is that Bullseye's future murders are hypothetical while his impending death is actual. Matt can prevent Bullseye's death now, and then devote himself to stopping him from killing anybody later. (This may be a naïve or futile plan, but Matt can be quite an idealist when he wants to.) Also, even though Matt's Catholicism has not been emphasized much at this point, it does teach that all life is not only valuable but redeemable, that all sin can be forgiven and even the most broken soul can be saved through confession and repentance. Even though Bullseye is unlikely to repent, Matt has to believe that this is possible, and as long as this is true, Matt cannot stand by and let him die—especially if, in case Bullseye does change his ways, he won't kill anymore.

Perhaps this is why Matt is so upset when Bullseye does return to kill again, telling him (as we saw earlier) that "I saved your life. And you went free. And you killed—again. I've been carrying that around inside me ever since, Bullseye. It hurts. It hurts a lot."[16] Matt also openly admits, both to Bullseye here and Manolis earlier, that he dearly wants to end Bullseye's life—which provides yet another reason not to do it, because he would never be sure if he did for the right reason or the wrong one. The last thing Matt wants is to become like Bullseye himself: As he told Spider-Man's foe Doctor Octopus before pulling a switch to save the villain from electrocuting himself, "I should let you fry, Octopus. But then I'd be no better than you."[17]

A Different Choice

Something changes, though, after Bullseye kills Elektra.[18] Following a long battle, Matt and Bullseye end up balanced on a wire between two buildings. Matt jumps up to unsettle Bullseye, who starts to fall, but Matt catches him. Bullseye yells, "No! You won't save me—not like before!" and lifts his free hand, containing the sai knife he killed Elektra with, towards Matt's wrist. In the next panel, we see Bullseye's hand dropping from Matt's, while Matt says, "You'll kill no one—ever again!"[19] Bullseye drops to the ground, and by the end of the issue we see him bandaged from head to toe in the hospital, his spine shattered, no feeling in his extremities and no power of speech, but full of hate for the man who put him there.

As drawn, it is left ambiguous whether Bullseye stabbed Matt in the wrist before Matt let go of Bullseye's hand. Even if Matt did drop him before he was stabbed, it could be seen as perfectly rational: The end re-sult was the same, but at least he saved his wrist by dropping him before he was stabbed. Furthermore, Matt's ominous statement as Bullseye plummets toward the ground can be interpreted as simple relief that a murderous threat is gone, whether or not he actively had a hand in elim-inating it. (No pun intended—or was it? Again, ambiguous!) Certainly, the next time they see each other, in the hospital where Matt told him the story of Chuckie, he declines to end Bullseye's life, but this could have been because Bullseye seemed incapacitated and seemed to pose no future threat.[20]

Nonetheless, Matt chooses a particularly dramatic way to not kill Bullseye: playing Russian Roulette with a .38 revolver, with Matt taking turns pulling the trigger at his own head and that of his paralyzed foe. After the first shot comes up empty for Bullseye, Matt goads him a bit, suggesting some responsibility for the killer's present state:

> Hope you don't mind me doing the honors for you like that, Bullseye. It's just that I don't want you to miss out on all the fun, and you…well, why be polite? You're paralyzed. Can't move a muscle in your whole body. Can't even talk. And that's too bad. Yeah…too bad I did it for you.

As he tells Bullseye the stories about Chuckie and Jack Murdock, Matt continues to play the game, and when there's only one chamber left in the gun to test, he points the gun at Bullseye and performs a monologue:

> When I fight you, and beat you, and know deep in my heart that I'm right in what I do…When I hate you and your kind so fiercely I could cry…When I can see that you are black and evil and have no right to live…When, at last, at long last, I've got you set squarely in my sights…And I smell your fear…and it is sweet to smell…When it comes to that one final, fatal act of ending you…

Then Matt pulls the trigger and all that's heard is a simple click, after which he says, "My gun has no bullets," showing that he had no intent of killing Bullseye after all—but he did let him know how much he dearly wants to, and that it is only his belief in rules, as represented by the story of his father hitting him when he was young, that stopped him.

Despite the severity of his injuries, Bullseye does recover, and several issues later he resumes his murderous spree. After Matt chases him to Japan (where he worked alongside Wolverine, as recounted above) and gets injured, he returns to New York, where the Black Widow finds him working out in his home gym, furious again about Bullseye. He tells her, "This thing with Bullseye has gone on too long! He's done too much—" When Natasha mentions Elektra, Matt reveals his intentions, adding, "He's killed a lot of people. He won't kill any more. I'll finish him—now, today."[21]

After beating up several guys in a bar, Matt learns from Turk that Bullseye is holed up in a deserted sports arena. He settles any doubts about his plan—as well as the truth about how Bullseye fell earlier—when he thinks to himself:

> Nobody around, nobody to interfere—when I'm killing Bullseye. *I failed the first time I tried.* I won't fail now. I'll make sure he's dead. Ever since that day I let him drop to what I thought was his death, I've been in torment. I've been crushed under some terrible, nameless burden. Tonight will be my deliverance.[22]

So he *didn't* just drop him out of rational self-preservation—he dropped him in the hope of killing him. The impending stabbing may have made him feel better about it for a while, even trick himself into thinking he didn't mean to do it, but this makes clear that he very much did. Matt decided to solve his trolley problem the most direct way possible—kill the one to save the many—and what's more, now that he knows he failed, he seems determined to do it again.[23]

When Matt gets to the arena, he remembers seeing his father fight there once—not as a boxer, but as a professional wrestler (in a very "familiar" costume, no less). After the fight, he found his father crying in the locker room, and he told Matt he took the job because he needed the money but was ashamed for making a fool of himself. Although he wasn't particularly proud of boxing, "it's what I do and I try to do it as best I can. I try to do it… honorably, that's the word I'm lookin' for." He felt he had disgraced himself as well as the sport of boxing, "and there ain't no money in the world that can pay for that." He finished by telling his son, "There's never a reason for not bein' what you are."[24]

This memory may have saved Bullseye's life that day, but it doesn't look this way at first. When the two start to fight, Bullseye flings the sai he used to kill Elektra at Matt, who catches it and throws it back, hitting Bullseye in the shoulder. He says, "Pretty good! Almost as good as I would have done it. Only I would have aimed for the heart," to which Matt replies, "I did. I missed."

After subduing Matt, Bullseye takes the opportunity to tell him how much he hates him for saving him from dying under the wheels of the subway train—but when Matt dropped him and shattered his spine, "that was okay." His greatest resentment, though, is for Matt's refusal to kill him in the hospital, when "you walked away without shooting. Treated me like dirt… like I wasn't worth killing." Bullseye deeply craves Daredevil's respect, and he hates that he appeared pitiful in the front of his arch enemy—so pitiful that he was not worth killing, especially after he tried to do it once before.

Next, Bullseye plays on Matt's greatest fear:

Well, I'm doing the same for you—letting you live. Only every time I do a job—every time I ice somebody—I'm gonna make sure

68

you hear about it. I know you, pal. I've been inside your head. And that's the worst thing I can do to you—just let you live knowing you're responsible for everything I do 'cause you didn't take your shot when you had it.

At this point, Matt loses it, trapping Bullseye in a choke hold—a common wrestling move, *wink wink*—but before he can choke his enemy to death, he sees a vision of Jack saying what he told him earlier: "There's never a reason for not bein' what you are." Matt releases Bullseye and says, "I'm no killer. The burden I've been feeling all these months…It's not that I didn't succeed in killing you. It's that I tried to. I'm not a killer."

After this, Matt is once again resolute against taking a life, although he does experience moments of weakness and doubt. As we'll describe in detail later, the Kingpin eventually learns Daredevil's secret identity and sets out to systemically destroy his life. When Matt realizes who's behind it, he thinks to himself:

It's the Kingpin. Somehow he found out that I'm Daredevil. He bribed and threatened everybody it took to destroy me. I've given this a lot of thought….I've got my strategy. I'm going to go to the Kingpin and I'm going to kill him…

But almost immediately he has a change of heart: "No. I won't kill him. I don't do that. I'll just beat him until he promises to give me my life back."[25] By this point, he has learned not to indulge his vengeful impulses for more than a moment, to remember "what you are"—which may be a violent man, but not a killer. Matthew may still be a long way from sainthood, but he's much closer than he would have been had he repeated the mistake of the night he dropped Bullseye.

[1] *Daredevil: The Man Without Fear* #2 (November 1993).

[2] *Daredevil: The Man Without Fear* #5 (February 1994).

[3] Ibid.

[4] This assumes, of course, a writer who is interested in him finding other ways to resolve the situation. (We've all seen *Man of Steel*, right?) By the way, if you are wondering why young Matt is so vicious here, when we saw earlier that it took him many years as Daredevil to reach this level of brutality, this story comes from Frank Miller's retelling of Daredevil's origin in the miniseries *Daredevil: The Man Without Fear* (1993-1994), which integrated many concepts he introduced to the canon, such as Elektra, Stick, and his conflict with Kingpin—as well as his more violent behavior.

[5] *Daredevil* #101 (July 1973).

[6] *Daredevil* #183 (June 1982).

[7] Ibid.

[8] Ibid.

[9] Ibid. (This rooftop debate was repeated in season 2 of the Netflix series.)

[10] Ibid.

[11] *Daredevil* #184 (July 1982).

[12] This may be as good a time as any to mention the many parallels between *Daredevil* and *Batman*, which has been noted in the comics themselves (for example, when a kid sees Daredevil swing by and shouts, "Look! It's Red Batman!" in *Daredevil*, vol. 3, #24, May 2013). This should not be surprising, considering the many writers who have written both heroes, including most notably Frank Miller on influential runs, plus Denny O'Neil, Ed Brubaker, and Chip Zdarsky (who recently wrote both at the same time).

[13] *Daredevil* #169 (March 1981); the rest of the events and quotes in this scene are from the same issue. Recall from our earlier discussion of the same issue that Matt invokes the same two concepts after he saves Bullseye from the subway car, then tells Manolis, "I wanted him to die, Nick. I detest what he does…what he is. But I'm not God—I'm not the law—and I'm not a murderer."

[14] For the original presentation of the trolley problem, see Philippa Foot, *Virtues and Vices* (Oxford: Oxford University Press, 2002), chapter 2, and Judith Jarvis Thompson, *Rights, Restitution, & Risk: Essays in Moral Theory* (Cambridge, MA: Harvard University Press, 1986), chapters 6 and 7.

[15] There has been a resurgence in discussion of the trolley problem in the last decade; for instance, see David Edmonds, *Would You Kill the Fat Man? The Trolley Problem* and *What Your Answer Tells Us about Right and Wrong* (Princeton, NJ: Princeton University Press, 2013) and Thomas Cathcart, *The Trolley Problem, or Would You Throw the Fat Guy Off the Bridge: A Philosophical Conundrum* (New York: Workman, 2013).

[16] *Daredevil* #172 (July 1981).

[17] *Daredevil* #165 (July 1980). Hmm, he pulled a switch when he could have walked away… trolley problems are everywhere!

[18] Don't worry—she gets better.

[19] *Daredevil* #181 (April 1982).

[20] *Daredevil* #191 (February 1983), which all the quotes and events from his story are from. (Incidentally, this was Frank Miller's last work on the title until he returned for "Born Again" storyline three years later.)

[21] *Daredevil* #199 (October 1983).

[22] *Daredevil* #200 (November 1983), emphasis mine.

[23] The situation with Bullseye is by means a standard trolley problem—for one thing, the sole person of the other track in the standard story wasn't responsible for putting the five passengers in danger! I explored many of these differences in the context of Batman and the Joker in *Batman and Ethics*, pp. 130-147.

[24] *Daredevil* #200 (November 1983), which the rest of the events and quotes about Bullseye are from.

[25] *Daredevil* #228 (March 1986).

Chapter 6: Value, Motivation, and Character

Value: Why Should He Be Daredevil?

As we've mentioned before, few heroes obsess over their value as a crimefighter and superhero more than Daredevil. Other heroes may beat themselves up on occasion for failing to save everybody (Spider-Man), not saving everybody quickly enough (Captain America), or not protecting the entire planet from all mystical threats (Doctor Strange). But none takes it as far as Matt Murdock, who repeatedly questions himself and whether he should continue as Daredevil. Peter Parker may have his famous "Spider-Man No More" cover (from July 1967's *Amazing Spider-Man* #50), where he leaves his costume in a garbage can and walks away, but there aren't enough garbage cans in Hell's Kitchen to hold his red tights every time Matt swears that he's "Daredevil no more."

We can look at a handful of episodes from his early years to see what Matt is so concerned about. These periodic episodes of doubt follow a similar pattern: At the beginning of the story, something makes Matt question what he's doing as Daredevil, but then by the end, he'll have helped or saved someone, which reminds him of the good he does.

As we described earlier, like most superheroes Matt is naturally concerned with the danger he puts his loved ones into. We already saw him say, while thinking about Karen, that "only by giving up my crime-fighting career—forever—could I marry Karen—without endangering her life!"[1] After Heather Glenn's father takes his own life, we also see that Matt wants to quit, as he thinks to himself that he "only wanted to help, but instead I failed everyone who ever meant anything at all to me!" Yet later, after saving some lives, he realizes that "maybe I have made my share to mistakes—both as Matt Murdock…and Daredevil…but I've done a lot of good, too! That may not justify the mistakes…but it does make them a lot easier to live with."[2]

Matt is also worried about the effects of his actions on people outside his narrow circle of friends and colleagues. Once, after he and the villain he's fighting almost destroy a family's home, Matt swears to quit, saying, "I once believed I became a swinging super-hero to help people. Instead, I'm destroying them.…For the first time in years, I can see things clearly. I wasn't helping anybody but number one!" Yet soon afterward he's back in action, acknowledging that "I've got special abilities, and because of that, I've got responsibilities—not only to myself, but to everyone."[3]

Sometimes Matt even considers how much being Daredevil prevents him from helping people more as a lawyer. After the grand opening of the Storefront, his free legal clinic, he thinks to himself (while in costume as Daredevil):

> There's a need for someone like Matt Murdock working to help those in need. And there's something in me which demands I do it. Maybe it's time that Daredevil wasn't the only hero in the Murdock family…because I've just given myself the chance to make Matt a greater hero than DD's ever been. Which only goes to prove that you needn't be super to be a hero…[4]

Nevertheless, after saving people as Daredevil, he elaborates: "I've questioned myself, wondered if I was doing the right thing by playing super-hero…but I guess as long as there are people who are forced to become victims—there's just got to be Daredevils to help them."[5]

Of course, Matt is only human, so he also doubts his future as Daredevil because of the costs it imposes on him personally. As we've seen, his superhero career often stands between him and a romantic future with Karen Page, both before and after she learns of his double life, as well as making trouble for him and his best friend Foggy. For instance, in an early moment of frustration, he says to himself, "Daredevil! He's already caused me to lose the girl I love—and my best friend, as well!" But after he runs into Willie Lincoln, a blind veteran he helped out in a previous story who had since gotten back on his feet, Matt realizes that "when I think of him, I figure maybe Daredevil's career wasn't all a waste! It's worth taking a dozen pratfalls for one guy like Willie Lincoln!"[6] Another time, he asks himself, "What would you do without your adventures to entertain you, Matthew, son?" He thinks of the life he could have had, full of marriage, kids, and free time, but realizes: "Sure. You might have had a better life— And maybe a few people who are alive because of DD—would be dead."[7]

All of these episodes leave the general impression that despite all the disadvantages to being Daredevil, including the harm it does to himself, his loved ones, and assorted other people, at the end of the day he does enough good to make up for it all, which he feels a responsibility to do. As he said in a passage quoted earlier,

> when I'm finally disgusted with being Daredevil, I remember why
> I am what I am. I remember all the people I've helped. And I know
> why I do what must be done. Sometimes you've got to do a job you
> hate because there's no other choice.[8]

This is consistent with his comment about responsibility quoted above, as well as other times he mentions it, such as when he told Heather, "I have certain responsibilities," before revealing the promise he made to his father to make something of himself, implying the promise itself created the responsibility to use his abilities for good.[9]

Another time, he suggests that the accident that gave him his powers also vested him with the responsibility to use them. Standing in front of a picture of Karen on a movie poster and resenting her for leaving him, Matt thinks to himself, "I've a responsibility, Karen—to myself, and the act of fate that gave me my powers—I can't betray that responsibility—not for you, God help me—not for anyone."[10] (If it seems strange to think he holds himself responsible to an act of fate, consider that he could be thinking of a personalized source of events.) Despite what we said in chapter 3 about his exaggerated sense of responsibility, it is an important source of motivation for Matt to keep going even when he doubts his value or contribution as Daredevil.

Motivation: Why Is He Daredevil?

Ironically, given Matt's comment about the job he hates, the reason for doubting his unique value is wondering if he just "plays" at being Daredevil because he loves it so much. Few other heroes enjoy what they do as much as the Man without Fear, who did not take long to earn his reputation as a "swashbuckler." In an early case, he intervenes when a group of men try to hijack a boat. Daredevil is cracking wise, jumping around and toying with the men, while saying things like, "No applause…please! I'm only doing what any talented, heroic…unhhh!" His final exclamation came as one of the hijackers shot him, after which he intentionally fell into the water, thinking that "this isn't the most glamorous exit scene." Reflecting on the episode later, he compares himself to Peter Pan, asking himself, "Do I really do this to help mankind…or am I just a showoff who never grew up?!!"[11]

Sometimes he takes his suspicion seriously enough to reconsider what he does altogether. When he wonders why he prefers being Daredevil to Matt Murdock, even if it means he can't be with Karen, his conclusion is that "I wanted the excitement—the glamour—the adventure of a superhero's life!"[12] During another similar episode, he asks himself, "Why do you do it, glory-seeker? Why?"[13]

But after he chastises himself as a "glory-seeker," Matt realizes, as usual, that he does good nonetheless. After one successful case that concludes with Daredevil returning a sick child to his mother, he thinks:

> Every so often I wonder if I'm doing the right thing playing super hero... if I'm really helping people, or just myself. But after a night like this, I don't have to question myself. The happiness I sensed in that mother's voice is enough for any man.[14]

Here, Matt acknowledges that the question of motivation is not simple and need not have one unique answer. In another story, while Daredevil works a case, a mysterious narrator plumbs his psyche for his motivation, and suggests the same ambiguity behind them:

> Even in those movements which seem so simple, so uncluttered with complexity...when good battles evil on the most obvious of terms...Even in these—the questions must be asked—the questions that determine the higher moral motive— And in this case— who can say what true motive moves Matt Murdock?...Why did you fight so hard?...Did you fight—out of altruism? Or—because of some less noble reason? Did you battle—out of some momentary exhilaration—or—because of pain—a pain now buried in your subconscious—the pain of Karen's seeming rejection? Answer, hero—before it's too late, and you never cease to wonder.[15]

The narrator need not worry about that—Matt Murdock will never cease to wonder why he does what he does!

For most of us, it isn't enough to simply do good, we also want to believe we are good—and to many philosophers, this depends on our motivations. As we saw in chapter 1, Immanuel Kant believed that to be a moral person, it isn't enough to act according to duty, but we must also act for the sake of duty itself. In other words, we should do the right thing for the right reasons. This makes sense: After all, when someone offers to help us but we suspect they have an ulterior motive, we think less of them (even if we accept the offer). With regard to superheroes, we expect them to help people for the sake of helping people, not to get good publicity, impress their crush, or get free coffee at their favorite shop. (Writers, on the other hand…well, never mind.)

When Matt worries that he's being Daredevil for the joy of it, it's a legitimate concern—except that it's doubtful he *only* does it for that reason. There's nothing saying you can't enjoy your job, even if your job involves risking your life on the daily to save lives. Even Kant acknowledged the existence of *mixed motivation*, or the possibility that we might have multiple reasons for doing the right thing, some of them explicitly moral (to save people) and some not (to have fun). Because of this, he wrote, no one can be certain they're doing anything for the right reason: "A human being cannot see into the depths of his own heart so as to be quite certain, in even a single action, of the purity of his moral intention and the sincerity of his disposition."[16]

So Matt may be right to ask whether he's only helping people as Daredevil for the thrill and glory, but as long as this isn't the only reason he does it—as long as he's also doing it for the sake of helping people—then he's doing fine. If he were only doing it for the kicks, he would stop saving lives once it loses its appeal. This was Kant's main concern with people who "find an inner pleasure in spreading joy around them and can rejoice in the satisfaction of others as their own work": If this were their only reason for behaving ethically, then when it stops, they would lose all motivation to do it.[17] In this case, we can be sure they weren't good people, despite the private joy they derived from doing good things—but it's safe to say this does not apply to Matt.

The Virtues of Daredevil

Our discussion of whether Matt Murdock is an ethically good person invokes not only the deontological ethics of Immanuel Kant but also the *virtue ethics* of the ancient Greeks and Romans such as Aristotle and the Stoics. Although there are differences in the fine details, most versions of virtue ethics focus on the same thing: the moral qualities or character traits of individuals themselves, rather than the things they do.

As we've seen, deontology and utilitarianism consider whether certain actions are ethical, such as lying, violence, or killing. Once this determination is made, this moral status can then be reflected back on the person doing them: Because lying is wrong, someone who lies would generally be considered a bad person (absent a valid justification). Virtue ethics turns this around and starts with the person: Dishonesty is a negative character trait, which leads a dishonest person to lie more often or more easily than an honest person would. This does not mean that an honest person never lies, only that they're less likely to, and when they do, they usually have a good reason for doing it (such as lying to spare someone pain). Like Kant, the virtue ethicists acknowledged human fallibility and imperfection, and they regarded positive and negative character traits—or virtues and vices, respectively—as forces that do not determine a person's choices, but merely influence them for better or worse.[18]

Most costumed crimefighters in comics, television, and movies possess the standard virtues we see in real-world heroes, such as courage, integrity, and a willingness to sacrifice for others. But not all superheroes (or real-life heroes, for that matter) make their moral decisions based solely on their virtues. We've seen that Matt prefers to follow rules whenever he can, which identifies him as closer to deontology than virtue ethics, while he's also concerned with consequences, which implies a utilitarian element to his moral thinking. Other heroes, such as Iron Man, base their thinking more explicitly on utilitarianism, trying to do the greatest amount of good for the greatest number of people while being less concerned with deontological issues of right and wrong—which creates a similar conflict when he does recognize them.[19] Among the Marvel heroes, Captain America is usually considered the epitome of virtue, standing up as a living symbol of integrity, but he also has a strong deontological streak: Once he's settled on the right thing to do, he follows that judgment without fail.[20]

Even though Matt doesn't often make his moral choices based primarily on his virtuous character traits, we can still identify the ones that are unique to him as a hero—and, while we're at it, the less virtuous ones as well. This will help us narrow down his *moral character*, which is another way to think of who Matt Murdock is—a question he asks himself an awful lot!

Naturally, the first virtue that comes to mind with respect to the Man without Fear is courage, a character trait he shares with most heroes (whether "super" or not). As it's usually understood, courage is not the absence of fear, but rather the ability to overcome and push past it when something important is at stake. Real-life heroes face storms, fires, and bullets to save lives, all without any superpowers. Matt does have powers, specifically his enhanced senses that allow him to detect both things to avoid (dangers) and things that need his attention (people to help). But these powers don't make him any faster, stronger, or invulnerable—his training with Stick may have enhanced Matt's abilities in these regards to some extent, but he's still no Thor, Luke Cage, or Spider-Man.

These considerations become relevant when we question whether Matt has the right "kind" of courage to truly count as virtuous, because it's not as simple as simply being courageous or not. According to Aristotle, usually regarded as the "main" virtue ethicist (in the same way that

Kant is consider the "main" deontologist), one needs to practice virtues such as courage in the appropriate way given the circumstances: "Whoever stands firm against the right things and fears the right things, for the right end, in the right way, at the right time, and is correspondingly confident, is the brave person."[21] This points out the importance of using judgment to determine the right way to practice virtue in a particular situation. For example, a trained firefighter can be expected to exhibit more courage than your average bystander when faced with a burning building. If the untrained bystander runs into a fire to save a child, their courage may seem noble, but it is also unreasonable and dangerous, more likely to get them hurt or worse (and probably put a firefighter's life at risk to save them).

Another concept of Aristotle's gives us a different way to think about the "right" way to practice a virtue such as courage. According to him, a virtue itself is found in the middle, or the "golden mean," between two extremes. For example, true courage is found between cowardice (insufficient courage) and foolhardiness (excessive courage). Again, judgment is required to find the golden mean in any given situation, which depends on the person and the circumstances they face, such as with the firefighter and bystander above: The courage shown by the firefighter would be foolhardy if the bystander did it, and if the firefighter acted like a bystander, they would be considered cowardly. But when the firefighter rushes into the fire while the bystander...well...stands by, they are each finding their own golden mean (or, in other words, practicing courage in their specific "right" way).

How does Matt's courage look in terms of these concepts? In one of his earliest stories, he faces off against Namor the Sub-Mariner, who clearly outclasses Matt in terms of strength, speed, and invulnerability (not to mention flight). During their first battle, Namor pulls Matt underwater while he tells him he respects his courage; after he passes out,

Namor quickly brings him to the surface, saying, "Such valor as yours does not deserve to perish!"[22] When they fight again, Matt exhausts all his resources fighting to prevent Namor from striking out against the surface world, and after he collapses on the ground, totally spent, Namor walks away, saying, "I have fought the Fantastic Four, the Avengers, and other super-powered humans, but none have been more courageous than he, the most vulnerable of all."[23] Despite the tenor of his praise, Namor's last words suggest that Matt's courage is excessive, far above the golden mean for a person with his level of strength compared to his enemy, a superpowered Atlantean. (But it did earn his respect, which is no small thing if you know Namor!)

Later comments from Matt himself reinforce this judgment about his level of courage, especially ones that reference his blindness. "I've said it before, and I'll no doubt say it again," he thinks to himself as he hears a distant cry for help. "If I could see what I'm doing…I'd be scared to death."[24] The idea that his lack of sight prevents him from appreciating danger is reinforced by the narration in a later issue as he jumps off a bridge:

> At the highest point of the span, he leaps wildly, recklessly into space. For another, the sprawling city below would promise certain death. But he has no view to trouble him. He's blind.[25]

Of course, we know that, although he is blind, his other senses (including his radar sense) compensate for this to some extent, especially regarding surfaces among which he runs and leaps, so we shouldn't take these comments too seriously. Nonetheless, they do reveal that Matt knows all too well how recklessly he behaves (consistent with his nickname).[26]

Besides courage, Matt's most frequently cited virtue is his perseverance. In a way, it was perseverance as much as courage that Namor admired so much in Matt: His courage helped him stand up to the Sub-Mariner in the first place, but it was his indefatigable resolve that kept him fighting until the last breath. Soon after he faced Namor, Matt fought Spider-Man—no slouch either—who later thought to himself about Daredevil: "I'll say one thing for him—he doesn't know how to quit!"[27]

Much later, when Matt struggles to open the massive door to the Kingpin's vault in search of evidence against other crime figures, he thinks to himself, "It's just too heavy...but there's so much at stake...I can't give up...and I won't." As the narration picks up from Matt's thoughts, it describes his steely resolve:

> It is hopeless. He knows. So he doesn't hope. He simply pulls—until his shoulder blades bulge against the knotted muscles of his back—until his breath hisses hotly through clenched teeth—until corded sinew stretches, near breaking—until his arms shake, and threaten to yank free of their sockets...and somehow, somewhere beyond the pain...he finds the strength he needs.[28]

After Matt flees from his disastrous experience at the Natchios estate discussed in chapter 1, there is similar exposition:

> Matt gets all the way back into town. He can't remember how. His senses come and go. He has no idea how many times he falls. He crawls deep inside himself. Through waves of nausea and clammy icy cold. He concentrates—concentrates—and he finds the strength he needs.[29]

It goes for a while longer, detailing how Matt makes it back to his dorm, repeating over and over, "Stay conscious...clean the wound...stop the bleeding," until eventually, "the bleeding stops and, light-headed, he knows he will live."

If you don't want to believe the mysterious narrator of this tale, take it from Matt Murdock himself. Throughout his early years, if not his entire career, his most formidable enemy in a physical fight has been Bullseye, who is not only an expert marksman with anything he can throw, fling, or flick, but an excellent hand-to-hand combatant as well. Often when they fight it is only Matt's perseverance that gives him the edge—and he usually makes sure Bullseye knows it.

During one fight, after Bullseye hits him in the head with a billiards ball and knocks out his radar sense, Matt is struggling and exhausted, but nonetheless tells his foe:

> Maybe I am half dead—maybe I should have laid down and died a long time ago—but I didn't! And I won't—ever! Remember this, Bullseye! Remember that Daredevil had the guts to go on and beat you after he should have given up and died! Because that's why Daredevil will always beat you![30]

Matt backs up his words when Bullseye shoots him in the shoulder and he still fights on. Next time they meet, Bullseye tries to shoot Matt again but can't stop shaking. When he says he's "not finished yet," Matt responds with a bluff: "Aren't you? Then go ahead, shoot. You tried that before, remember? It didn't stop me then. And it won't stop me now. Nothing you can do will stop me now."[31] This is too much for Bullseye's fragile psyche and he cracks, later revealed to be the effect of a fatal brain tumor. Finally, in the infamous scene in the subway tunnel, before an exhausted Matt defeats Bullseye (and then saves his life), he tells him, "I told you once, Bullseye…a long time ago…I never give up…that's why…I'll always beat you…"[32]

As with courage, Matt's perseverance is almost superhuman—or approximates what a superhuman's perseverance might look like. Is it also excessive, though? In the example cited above, Matt definitely pushes himself to his absolute limits, especially when he believes his cause is just. But he doesn't exceed his limits—he can't, by definition, because he can only do what he can do. His perseverance, as extreme as it might seem, has to respect his physical and mental limits; it can't surpass them, as his excessive courage does, pushing him to face threats that are unreasonably challenging. Ideally, an appropriate level of courage would help Matt judge better which foes to fight and which to deal with in other ways, and his perseverance would help him carry through whichever decision his courage helped him to make. Although his perseverance is impressive given his lack of super-strength or invulnerability, it also helps compensate for these limitations, in the way that boxers with stamina can pace themselves and outlast a stronger, faster opponent.

The Most Important Vice of Daredevil

We'll return to perseverance later when we discuss the culmination of this era of Daredevil's stories, but now we should address one of Matt Murdock's less virtuous character traits: his lack of self-control. It isn't just the overwrought dialogue, especially in the early stories, that makes Matt stand out as a particularly dramatic figure in the Marvel Universe. It's his passion, which can be a great thing, driving him to do the most he can to help the people around him, both as a lawyer and a superhero. But when it pushes him to make rash decisions in an emotionally hot state—especially when he's in the throes of romantic turmoil—it compromises the degree to which he can be an effective hero in either role.

Self-control and the lack thereof was a popular topic among the ancient virtue ethicists. Aristotle wrote in terms of both *temperance*, which deals with resisting excessive desires and bodily pleasures, as well as *incontinence,* which is more about doing things you know are wrong or harmful (or what we today call "weakness of will").[33] Perhaps the Stoics are best known for this, as most of them emphasized in some way the need to control one's impulses and desires in order to make more rational decisions. Today, the word "stoic" describes an unemotional, rational person, which is a bit of an exaggeration: The Stoics did not recommend the outright elimination of emotion and desire, but rather putting them in their proper context in which they influence but do not rule over one's choices.[34]

To be fair, although Matt is prone to lose control and act rashly, he's also self-aware about this. For example, he remembers the woman whose death he recklessly caused while going after his father's killers; later, after being stopped by police while chasing Elektra in the park, his impulse is to fight and escape, but he resists. As the narration reads, "He remembers the last time he lost control. He remembers shattering window glass. He remembers a pathetic prayer to God."[35] But this does not last: Not long afterward, he viciously beats some muggers, working out his childhood trauma over being teased and taunted for studying so much and refusing to fight back because of the promise he made to his father. Then, he retreats to his father's old gym, at which point the exposition reads, "His fever breaks. He's in control again."[36]

We've seen this excessively emotional side to Matt throughout our exploration of his early adventures. His anger was on display particularly in his many battles with Bullseye, who more than any foe tests Matt's self-control. Matt has to be at his absolute best to defeat Bullseye, who usually pushes him to the brink of death himself, but as much as he wants to Matt will not let himself kill the homicidal psychopath. This forces him to walk a fine line, indulging his anger enough to drive him to the peak of fighting ability, but not too much to push him over the edge into murder.

The fact that he usually resists this urge is a good sign, showing that, when it counts the most, Matt *can* control his worst impulses, even when his physical and mental resources are completely drained.[37] This corresponds to Aristotle's views of anger, which found it useful in moderation, using similar language to what he used with reference to courage: "The person who is angry at the right things and towards the right people, and also in the right way, at the right time and for the right length of time, is praised."[38] The appropriate use of anger can prompt us to action against injustice, whether with a protest sign and a bullhorn or a red costume and a billy club. The Stoic philosopher Seneca disagreed, however, believing that anger is a particularly insidious emotion: "It is easier to exclude the forces of ruin than to govern them… once they have established possession, they prove to be more powerful than their governor, refusing to be cut back or reduced."[39] Throughout his career—including past the early period we're surveying here—Matt continually tests which understanding is correct.

[1] *Daredevil* #43 (August 1968).

[2] *Daredevil* #151 (March 1978).

[3] *Daredevil* #128 (December 1975). (Does Peter Parker get a nickel every time someone says something like this?)

[4] *Daredevil* #130 (February 1976).

[5] Ibid. This flies in the face of his earlier statement that "Do-gooders are a glut on the market these days" (*Daredevil* #49, February 1969), but to be fair, his foe was hypnotizing people visually, so Daredevil was uniquely able to defeat him. (Because, you know, no one else could close their eyes or anything.)

[6] *Daredevil* #49 (February 1969).

[7] *Daredevil* #78 (July 1971).

[8] Ibid.

[9] *Daredevil* #160 (September 1979).

[10] *Daredevil* #78 (July 1971).

[11] *Daredevil* #9 (August 1965). Later, as an older couple see Daredevil swings by, Henry says, "Look at him, swinging up there like he was king of the world. I tell you, Martha—J. Jonah Jameson is right! Those super heroes are a nuisance to this city... and they should be run out on a rail!" Martha replies, "You better believe it, Henry. They're only interested in themselves... What do they ever do for us?" (*Daredevil* #139, November 1976). They were wrong in this case, but we can assume that Matt would nod in agreement to the general sentiment.

[12] *Daredevil* #49 (February 1969).

[13] *Daredevil* #75 (April 1971).

[14] *Daredevil* #139 (November 1976).

[15] *Daredevil* #80 (September 1971).

[16] Immanuel Kant, *The Metaphysics of Morals*, trans. and ed. Mary Gregor (Cambridge: Cambridge University Press, 1797/1996), p. 392.

[17] Immanuel Kant, *Grounding for the Metaphysics of Morals*, trans. James W. Ellington (Indianapolis, IN: Hackett Publishing, 1785/1993), p. 398. This passage has led many people to conclude incorrectly that one needs to be miserable to be moral, which we *definitely* do not want Matt to hear!

[18] For a survey of virtue ethics, see Rosalind Hursthouse and Glen Pettigrove, "Virtue Ethics," *The Stanford Encyclopedia of Philosophy* (Winter 2022 Edition), Edward N. Zalta & Uri Nodelman (eds.), at https://plato.stanford.edu/archives/win2022/entries/ethics-virtue/.

[19] Until I write an entire book on Iron Man, see Part I of my book *A Philosopher Reads... Marvel Comics' Civil War: Exploring the Moral Judgment of Captain America, Iron Man, and Spider-Man* (Ockham Publishing, 2016).

[20] See my book *The Virtues of Captain America: Modern-Day Lessons on Character from a World War II Superhero* (Hoboken, NJ: John Wiley & Sons, 2014), pp. 13–20.

[21] Aristotle, *Nicomachean Ethics*, trans. Terence Irwin (Indianapolis, IN: Hackett Publishing, 1985), p. 1115b.

[22] *Daredevil* #7 (April 1965).

[23] Ibid.

[24] *Daredevil* #78 (July 1971).

[25] *Daredevil* #170 (May 1981). (The comic reads "sprawing," but I assume it should have said "sprawling.")

[26] For more on to the pervasive belief that his other senses "make up" for his blindness, see Christine Hanefalk's post "'My other senses more than compensate'" at her blog *The Other Murdock Papers* (https://www.theothermurdockpapers.com/2008/06/my-other-senses-more-than-compensate/), and her book *Being Matt Murdock: One Fan's Journey into the Science of Daredevil* (TOMP Press, 2022).

[27] *Daredevil* #17 (June 1966).

[28] *Daredevil* #171 (June 1981).

[29] *Daredevil: The Man Without Fear* #3 (December 1993).

[30] *Daredevil* #146 (June 1977).

[31] *Daredevil* #161 (November 1979).

[32] *Daredevil* #169 (March 1981).

[33] Aristotle, *Nicomachean Ethics*, 1117b–1119b (on temperance) and 1145a–1152a (on incontinence).

[34] For readable introductions to Stoic thought—which lends itself very well to a self-help context—see William B. Irvine, *A Guide to the Good Life: The Ancient Art of Joy* (Oxford: Oxford University Press, 2008), and Massimo Pigliucci, *How to Be a Stoic* (New York: Basic Books, 2018).

[35] *Daredevil: The Man Without Fear* #2 (November 1993). The one time he didn't resist, and let Bullseye fall to his possible death, he seemed calm and rational, as described above.

[36] *Daredevil: The Man Without Fear* #4 (January 1994).

[37] On the theory that willpower is like a muscle, increasing when used and decreasing when exhausted or neglected, see Roy F. Baumeister and John Tierney, *Willpower: Why Self-Control Is the Secret to Success* (New York: Penguin, 2011).

[38] Aristotle, *Nicomachean Ethics*, p. 1125b30. For a contemporary application, see Myisha Cherry, *The Case for Rage: Why Anger Is Essential to Anti-Racist Struggle* (Oxford: Oxford University Press, 2021).

[39] Seneca, "On Anger," collected in John M. Cooper and J.F. Procopé (eds), *Seneca: Moral and Political Essays* (Cambridge: Cambridge University Press, 1995), p. 25.

Chapter 7: The Fall and Rise of Daredevil

To Be or Not to Be Daredevil

When we think of Matt Murdock's issues with self-control, his penchant for violence might be the most visceral example, but it is his all-too-human desire for a successful love life that tests his resolve more than anything. We see this reach its peak after Elektra's murder, when Matt finds her grave empty and learns that the Hand is trying to resurrect her and enlist her to their cause. Stick and his allies want to prevent the resurrection, but they suspect Matt will interfere in the hopes that he can be reunited with her. As Stone, another one of Stick's students, describes Matt to the Black Widow while they search for Elektra's corpse, "The adventurer cannot control his passions. Emotions are dangerous to us, in this struggle. I fear his may be hot enough…to make him betray us."[1] In this case, Matt's passion threatens to contribute to the Hand's evil agenda, perhaps the most extreme case of his heart ruling over his head.

A related example of Matt's indecisiveness is his repeated back-and-forth over whether he will continue being Daredevil, which is often due to wanting a more fulfilling personal life that includes a romantic relationship. We saw several of these episodes earlier when we discussed his internal struggles with purpose, which is how these episodes usually end: He swears to give up being Daredevil so he can be with Karen or Heather, but soon realizes Daredevil's unique value as a hero and decides to carry on.[2]

We see this for the first time early in his career after a convoluted episode in which Matt concocts a brother "Mike," reveals him to be

Daredevil, and then fakes both their deaths.[3] Now enjoying the simple life with Karen, Matt thinks to himself, "I'm beginning to think I should have killed off my Daredevil identity long ago!" But after the threat of the Jester compels him to don his costume again, he realizes that "so long as Matt Murdock lives…there will always be…a Daredevil!"[4]

Seven issues later, though, after alienating both Foggy and Karen, Matt proclaims (to no one):

> I've had it! It's over! I'm giving up the role of Daredevil—forever! Daredevil! He's already caused me to lose the girl I love—and my best friend, as well! It's something I should have done—a long time ago![5]

He also curses the costume as a symbol of his double life, acknowledging the work he's put into making it "a symbol of courage and skill" before bemoaning what it's cost him:

> Big deal! It can't put its arms around me—or kiss me goodnight! Or set me on fire, like the sound of Karen's voice—or the touch of her lips! All it can do—all it's ever done—is keep me from her! But not any longer![6]

Nonetheless, as we saw above, by the end of the issue he comes to appreciate the value of Daredevil (after seeing Willie Lincoln again) and rescinds his resignation.

Matt promises to quit again literally two issues later—telling both Karen and Foggy he has an important announcement, which he never delivers—but another two issues later he introduces a twist, motivated by the discovery of his secret identity by the villain Starr Saxon.[7] After reviewing his own origin and early days, Matt remembers how he got so used to being called Daredevil that "I often wondered whether Matt Murdock was real—or just a product of Daredevil's imagination!" This leads to the realization that "my problem isn't Daredevil—and never was! It was always Matt—the blind lawyer—the hapless, helpless invalid! He's been my plague… since the day I first donned a costume!" Finally, Matt shouts, "Then, let Matt Murdock no longer exist!!"[8]

To his credit, he does follow through on this, faking Matt Murdock's death in a plane crash with the intention of reviving him after he catches Saxon. Well, we should give him partial credit: Saxon falls to his death, solving Matt's "problem," but Karen's father also dies soon thereafter. Matt chooses his funeral as the appropriate time to bring Matt Murdock back by approaching Karen as Daredevil and then unmasking.[9] As we saw earlier, he proposes to her in the next issue and promises to give up Daredevil to secure her "yes," but recants by the end.

Identity Crisis

Although this flip-flopping over giving up one of his two identities serves as an example of Matt's incontinence, or lack of conviction to a cause he knows in his heart is right and just, we can also pull back from the soap-opera drama to look at the more general issues with identity that it reveals—issues that contribute to the mental instability that will come to a head as we approach the exciting conclusion of his early years.

We saw above that, at one point, Matt asserts that "so long as Matt Murdock lives…there will always be…a Daredevil!"[10] But at other times he struggles with being both lawyer and superhero and identifying which he "really" is. When Karen asks him to quit being Daredevil so they can be together, he tells her, "Until I'm sure whether the real 'me' is Matt Murdock…or Daredevil…I've got to be both men!"[11]

It may seem natural to assume that Matt Murdock is the "real" identity, and Daredevil is the disguise—after all, Matt was there first! But our feelings about who we are change over the years, which is an issue of what philosophers call our "personal identity," or what determines and identifies who we are as individuals.[12]

Our personal identity distinguishes us from other people and also from who we are (or were) at different times in our lives. Are you the same person you were yesterday? It may feel like it because we change gradually and there is continuity in who we are from day to day. But just as the ancient philosopher Heraclitus said, you never step into the same river twice because it is always moving and changing, and you are not exactly the same "you" from one day to the next.[13] This effect is usually more noticeable over years than over days: Even if you feel like the same person you were yesterday, do you feel like the same person you were five, ten, or twenty years ago? Some people never realize this, though, feeling that the person they are deep down is fairly constant over time, while others have an epiphany at some point, suddenly feeling like they're living the wrong life, whether it's working at the wrong job, living with the wrong person, or living a gender they don't feel is right for them.

Matt's issue is distinct from these, as he is literally living two lives. His situation could be compared more dramatically to the person who maintains two families in different cities, but it might be familiar to anyone who feels like their work and home lives are completely separate, equally demanding, and impossible to reconcile. Appropriately, one time he leaves the office for his other "job," he says to himself as he changes clothes:

> I felt like I was suffocating in that business suit of mine! I'd have jumped out of my skin if I had to wait any longer to get into my working clothes! I never realized Daredevil was so much a part of me! It's like DD is my real identity—and I'm just play-acting as Matt Murdock![14]

Not long after, while he mulls quitting for Karen's sake, he admits that "sometimes I think I was born to be Daredevil—and, Matt Murdock is the identity that's not real!"[15] To some extent, most superheroes experience this identity crisis at one point in their lives, but with Matt it gets wrapped up in his ongoing struggle over quitting, compounding both conflicts.

Loneliness and Solitude

If that weren't enough, also tied into all of this is his persistent feeling of loneliness. Once again, this is hardly unique to Matt Murdock, but it seems to hit him even harder because of his strong desire for romantic companionship (which many other superheroes seem to be able to put out of their minds more easily). We've already seen several examples where he curses his solitary existence after losing Karen or Heather, acknowledging that his costume "can't put its arms around me—or kiss me goodnight" and that his responsibility "makes me the loneliest man in the world" and "is why DD walks alone."[16]

Almost from his first appearance, Matt realizes he loves Karen, but he also admits that she deserves more than a life with a superhero. When he learns that Foggy loves her too, he thinks to himself, "This must be fate's way of telling me that I'm destined to always be... a loner." After getting into costume, he dramatically proclaims, "Where Daredevil walks, he must walk alone! Thus do I accept my lonely fate!"[17] After considering telling Karen and Foggy his secret but quickly reconsidering out of concern for their safety, he once again says, "Matt Murdock has always been a loner... and a loner he must stay—so long as Daredevil lives!!"[18] (Note the condition at the end: Our clever attorney gave himself a way out, if he'd only use it!)

Matt's loneliness is not only romantic in nature, though; it goes much deeper than that, affecting his ability to confide in anyone about his true life. Although more people in his life gradually learn his secret identity—even before it's revealed in the public much later—in the early years very few know that Matt Murdock, attorney-at-law, is actually Daredevil, the Man without Fear. Other than the costumed colleagues who eventually pick up on it (as well as the occasional villain), Karen and Heather are actually the only two to know for many years—but as we'll see, this is little help to Matt.

We see this broader sense of loneliness when Matt, drowning in self-pity over his latest problems, tells himself:

> Stop complaining, DD. You're only doing it to hear someone talk. That's part of the trouble with this game—the blasted loneliness! You've been a towel shoulder for lots of people in your time, lawyer-man. Who do *you* turn to? Who *can* you turn to? You've alienated just about everyone who's ever tried to help you—Karen, Foggy—all of them.[19]

Later, in the scene at the grave of Heather's father, after Matt cites his responsibilities as the reason he can't attend to her and her family as much as she wants, she slaps him and leaves. He overhears Foggy tell their new secretary Becky Blake, who wants to go to him, that "I think he'd rather be alone." Later, Matt thinks to himself, "I hate being alone," then sympathizes with Heather's concerns before reiterating, nonetheless, that "sometimes I just get so blamed lonely... I need somebody to talk to. Someone who can understand what I'm going through."[20] Even though Karen and Heather are uniquely positioned to comfort Matt, as friends if not lovers, their own problems with Daredevil render them unavailable, leaving Matt without anyone to truly confide in.

From Joy to Despair and Back Again

Given this persistent sense of loneliness, as well as his feelings of failure and the pressure of having to serve two roles, each with its myriad conflicts, we might expect Matt Murdock to be sad and morose, if not depressed—and he often is, as we've seen throughout this book. Furthermore, this is not uncommon among superheroes, who are usually burdened by incredible responsibility and feelings that, as much as they do, they haven't done enough, especially when obligations to the public conflict with commitments to their family and friends.

Not only does this hit Matt harder than most, which is no surprise, but his sadness is also contrasted with periods of exuberant joy. As we noted above, Daredevil is the rare superhero who truly enjoys being a superhero. In an early issue, after he returns to New York City from an overseas adventure, he swings around the city, saying:

> How I've longed for this marvelous moment—to be back in action again—free as a falcon, with the world at my feet! This is where I belong! This is my food, my drink, my life! This is why Daredevil was born![21]

The particular elation expressed here may reflect the fact that Matt has to practice extraordinary restraint in his "civilian" life, hiding his enhanced senses as well as his athleticism, so his time as Daredevil allows him a unique freedom and release.[22] A similar scene is repeated several times in the first couple decades of his book, such as much later, when running through the city, he thinks to himself, "What a gorgeous morning! Wind in my face...summer sun at my back...all the endless city sounds like a brass band below me...You're beautiful, New York. I love you."[23]

However, many of his most enthusiastic expressions of happiness come very quickly after intense periods of sadness. For example, after

96

Matt fakes his death to thwart the blackmailer Starr Saxon, who then falls to his death, it is understandable that Matt would be relieved, but he's positively ecstatic. In the next issue, onlookers can't believe their eyes when Daredevil swings overhead while…well, as one person puts it, "Daredevil is—singing!!"[24] Later, after the tense episode with Bullseye on the subway tracks, Matt is seen running through the city with a huge smile on his face, joking with strangers, before visiting Josie's Bar to rough up Turk for information.[25] In the next issue, after Bullseye kicks him out the window of a skyscraper, we see Matt sitting in the park with Heather, the two of them dancing like Fred Astaire and Ginger Rogers.[26]

There are two ways to look at this behavior. On the bright side, he could be balancing the more difficult parts of his life with moments of joy and abandon, which would be healthy. But he could just as well be cycling from extreme lows to extreme highs, neither of them moderating or converging over time.[27]

The second interpretation is supported by one of the most disturbing episodes in Daredevil's early years. After Karen has apparently left him for good to be a Hollywood actress, Matt seems to have found serenity, thinking to himself, "Peace, DD. Just a few minutes…of peace. Can you dig it, hero-man? No worries…not a one! It's been the kind of day—that makes you forget all the others."[28] He begins to wonder if he ever really loved Karen, then catches himself and thinks, "Forget it! Matthew, old son, you're feeling good—and it's a feeling you want to feed." He seems grateful for this good mood while trying not to question it: "Maybe it won't last! Maybe in an hour, I'll be bleeding again. Sure. Maybe in an hour I'll be dead…Point being: Matthew, you just don't care! You're happy. You're alive. And you're livin'!"

As revealed by the narration later in the issue—the same issue with the long passages about motivation we discussed earlier—his exuberance was not only short-lived but illusory:

Your euphoria—your joy in living—all a show! All a self-deluding show! You know, don't you, Matthew? Your happiness was a front, to keep you from thinking of your pain. But a moment's delusion—to wipe out the constant memory of Karen, the girl you love.

He comes to this realization while battling the Owl, who leaves Daredevil hanging from a damaged helicopter that explodes before plunging into the Hudson River—all of which Karen watches on television.[29]

Even worse than the possibility that Matt may have died is the insinuation of the accompanying narration that he might have too easily accepted his fate, taking the opportunity for "a final escape" to end his pain once and for all:

All day—you've been escaping, Daredevil. First into a false joy…and then into battle. Escaping—from a vision of a love that is no more…! And now you're making the final escape…an escape even you can't survive…In a moment, it'll be over for you, Matt Murdock. Over for one man. Over for a small part of this greater tragedy—

This impression is reinforced at the beginning of the next issue, when Daredevil is shown sinking into the water, pondering his regrets, reconsidering what he could have done differently—but not making the slightest effort to swim or survive, until he eventually loses consciousness.[30] The Black Widow saves him just in time, but we are left to wonder: Did he really want to be saved?

Nonetheless, Daredevil springs back into action by the end of the issue, working with Natasha to defeat the Owl. At the opening of the next issue, he is back to swinging around the city, taking risks and barely surviving them, all the time saying things like "what's life for, if not high living?"[31] If Matt Murdock had achieved a healthy balance between happiness and sadness, ups and downs, you would think he would take more time to recover after a passive suicide attempt—but instead, the pendulum swings all the way to the other side, as it often does.

Born Again

We see the pendulum take its most drastic swing to this point, threatening to break loose altogether, in the epic "Born Again" storyline that caps our coverage of Daredevil's early years. This legendary tale gets underway when Karen Page resurfaces, down on her luck and addicted to drugs, and sells her own possession of value, Daredevil's secret identity, to a dealer in exchange for a hit. This priceless information eventually makes its way to the Kingpin, who begins to tear apart Matt's life—which was already in a bad state due to a deteriorating friendship with Foggy and his ex-girlfriend Heather's recent death by suicide.[32]

It is the latter event that helps send Matt into a depressive spiral. Several issues before "Born Again" starts in earnest, Matt visits Heather's grave and takes the opportunity to survey the state of his life, thinking to himself:

> Lonely graveyard on a chilly Sunday morning…a lot like my life, lately. Heather's death…the problems I've been having with Foggy Nelson…the uncertainty of my new relationship with my new love Glorianna…Not much to sing about. What was that saying I heard the other day? Life is hard and then you die.[33]

Just then, he catches Spider-Man's foe the Vulture robbing a nearby grave, and the two fight (as they do). When Matt accuses Vulture of having a "diseased mind," the villain responds that "I wouldn't go pointing fingers, boy. I get the feelin' you ain't exactly healthy yourself," and after he lifts Matt into the air, says, "You reek of death. Might be time you sampled some."

Given his pensive mood, Matt takes the Vulture's words to heart, asking himself if he has a death wish, especially considering how easily he let himself be swept up into the air by the villain. The Vulture wonders the same thing, asking Matt, "Why don't you give up, boy? You want to lose! Otherwise, it wouldn't have been so easy knocking you off that roof!" After they fight some more, Matt becomes angry, telling the Vulture, "A while ago, you said I secretly wanted to die. You were wrong." He then uses his fists to prove it

> to you and to myself—by beating you...You—and everything you represent—the death and decay that eat away at a man until he surrenders...the horror that pulls you down into the pit! Well, I'm not the surrendering kind, mister! Got that? I never give up!

This reminds us—and possibly Matt as well—of his perseverance, discussed earlier, and also of the anger that often accompanies it due to being pushed so very far. (His perseverance will prove invaluable once the Kingpin launches his plans, although he will need to control the anger that comes with it.)

Matt's statement at Heather's grave reflects another recurring theme in his life, one that will only be magnified by the Kingpin: the feeling that everything is lost. We've seen this a number of times, especially when he wants to quit being Daredevil and he expresses his frustration at the state of his life. For instance, after he tells Heather about her father's suicide and she throws him out, he thinks to himself, "My whole world's falling apart...and there's nothing I can do to stop it!"[34] Much earlier, after Daredevil is exonerated from a murder charge, onlookers say to each other that he must be happy, but all he can think about is how much he let down Foggy and lost Karen, "the most important thing in my life...forever," concluding, "Yeah—if I was any happier—I couldn't bear it!"[35]

The issue after the scene at Heather's grave, Matt catches himself thinking that he's in good shape "for his age," but then remembers he's not even thirty yet. After reflecting on the accident that stole his sight and granted him his powers, he adds that it was just the beginning of his bad fortune, just

> one disaster after another, and everyone putting the blame on me. Everyone I loved or trusted. Foggy—my partner—couldn't pick up just a little slack and keep the law firm going and Heather—yeah, Heather—why be afraid to think of her name?—Killed herself…this on top of everything else I have to deal with…but that's what life has turned into. One thing after another to deal with.[36]

Statements like this may suggest *apocalyptic thinking*, a psychological term for the tendency to exaggerate negative life events and assume they'll only get worse, but to be fair, Matt's life is going downhill fairly quickly at this point—and, in the issues to come, it will only get worse.[37]

"Born Again" pushes this theme to its limit as the Kingpin uses his knowledge of Daredevil's secret identity to systematically dismantle Matt Murdock's life, including destroying his finances and rigging criminal charges against him. Eventually, all Matt has left is Daredevil, about which he thinks, at least "that much I've done right with my life."[38] In his current state of fatigue-induced paranoia, also seen in the quote above, Matt suspects everyone in his life of conspiring against him, until he returns to his home only to watch it explode. After that happens, he realizes who is behind it all: "It was a nice piece of work, Kingpin. You shouldn't have signed it."[39]

Even after he's able to identify an enemy as the cause of much of his current troubles, Matt continues to decline into anger and self-pity. The morning after he loses his home, Matt wakes up in a run-down hotel, reflecting on the fact that he used to be a respected lawyer and admired

101

superhero, and dwelling on his lost sight (despite his other senses being greatly enhanced, which he acknowledges as well):

> Now I'm just a blind man…a blind man who's lost his job, his livelihood, his home, his girl…who fate gave the ability to hear and smell and taste better than anybody in the world can—which is a great way to catch all the misery of being alive.[40]

His anger comes through in his failure to restrain his worst violent impulses, choking the hotel manager (whom Matt thinks was sent by the Kingpin) and savagely beating up muggers on the subway and a police officer afterward—an event reported by one of the Kingpin's thugs as the signs of "a man possessed."[41] Matt even finds the Kingpin himself and attacks him, but stands no chance in his physically and mentally depleted state, so the Kingpin puts his body in a taxicab, beats the driver to death with one of Matt's billy clubs, and dumps the cab in the East River.[42]

But Matt Murdock does not die, his perseverance being tested like never before, and eventually he finds himself in a church, being tended to by Sister Maggie, whom he comes to realize is his long lost mother (although she denies it, to no avail).[43] As he trains himself back to fighting shape, Matt endures a series of Herculean labors, including facing an imitation Daredevil and a drugged-up super-soldier, both sent by the Kingpin to do what he could not do himself: kill Daredevil.[44] Despite the Kingpin's best efforts, Matt not only survives, but by the end of the story he also brings to light all of the Kingpin's criminal machinations—with the help of reporter Ben Urich, who also goes through it in this storyline—and, more important, reunites with Karen Page.

Although revenge against the Kingpin may have satisfied his concern for justice, it is finding Karen once again that brings Matt Murdock back from the brink. After Matt saves her from the fake Daredevil, she tells

him how she sold his secret for heroin, which later cost him everything—
or so she assumes. According to the narration, after she confesses, he
simply comforts her:

> "I've lost nothing," Matt said, and laughed like a boy—and Karen
> didn't understand—and Matt kissed her—and held her...[45]

The final page of the last issue of "Born Again" shows Matt and Karen
walking hand-in-hand down the sidewalk on a sunny day and smiling.
His thoughts reiterate how far he's come since his self-pitying thoughts
at the beginning of many recent issues:

> My name is Matt Murdock. I was blinded by radiation. My re-
> maining senses function with superhuman sharpness. I live in
> Hell's Kitchen and do my best to keep it clean. That's all you need
> to know.[46]

No more does Matt complain about how much he's lost or how everything has gone wrong for him. Nothing has improved in material
terms, but after what he has endured and survived, he knows now that
he can overcome anything. Bad things will continue to happen—and, to
be sure, they do—but he cannot make them better by going on about
how bad his life is. It is his life to start anew, and no one can do this but
him. In modern terms, the Kingpin has given Matt an opportunity to
reboot his life as Matt Murdock, former lawyer, which he does with Karen at his side—and with Daredevil still lingering in the background,
ready to come between them as he always has.

Although Karen's return to his life played a large role in Matt's "re-
birth," I think more credit should go to his victory over the Kingpin,
which took all of his resolve to survive and his refusal to be beaten. One
of the most powerful individuals in the city, whose reach extends into
the worlds of business, finance, and government, as well as the criminal
underground, levied all his influence to attack Matt Murdock from every

angle—and still, he could not beat him. Fisk took everything away from Matt, not only his worldly possessions but his reputation as well, yet he could not take away who Matt Murdock is—his belief in right and wrong, his moral character, and his unquenchable drive to fight and survive.

What's more, at his lowest point Matt won his final victory against the Kingpin without throwing a punch—fulfilling his promise to his father, at least when it mattered most.

[1] *Daredevil* #190 (January 1983).

[2] Sometimes, though, it's resolved when he remembers how much he loves being Daredevil, as we discussed earlier. One of the many times he asked himself why he doesn't quit for Karen, he answers, "There's something about being a free-wheeling superhero... something that gets in your blood!" (*Daredevil* #61, February 1971). But two issues later, he changes his mind, thinking, "Karen's more important to me than life itself—let alone my double-life as a swashbuckling showoff!" (*Daredevil* #63, April 1970).

[3] "Mike" first appears in *Daredevil* #25 (February 1967) and "dies" in *Daredevil* #41 (June 1968). (Surprisingly, he "came back" for real in *Daredevil*, vol. 5, #606, October 2018.)

[4] *Daredevil* #42 (July 1968).

[5] *Daredevil* #49 (February 1969).

[6] Ibid.

[7] And I didn't even mention the time, recounted above, that he swore to quit because he felt he was doing more harm than good but soon realized he has responsibilities he cannot ignore (*Daredevil* #128, December 1975).

[8] *Daredevil* #53 (June 1969).

[9] *Daredevil* #57 (October 1969).

[10] *Daredevil* #42 (July 1968).

[11] *Daredevil* #61 (February 1971).

[12] For an approachable introduction to the philosophy of personal identity (and a unique perspective on it), see Kathleen Wallace, "You Are a Network," *Aeon*, May 18, 2021, at https://aeon.co/essays/the-self-is-not-singular-but-a-fluid-network-of-identities. For more detail, see Eric T. Olson, "Personal Identity," *The Stanford Encyclopedia of Philosophy* (Summer 2022 Edition), Edward N. Zalta (ed.), at https://plato.stanford.edu/archives/sum2022/entries/identity-personal/.

[13] On this, see section 3.1 in Daniel W. Graham, "Heraclitus," *The Stanford Encyclopedia of Philosophy* (Summer 2021 Edition), Edward N. Zalta (ed.), at https://plato.stanford.edu/archives/sum2021/entries/heraclitus/.

[14] *Daredevil* #25 (February 1967).

[15] *Daredevil* #43 (August 1968).

[16] *Daredevil* #49 (February 1969); *Daredevil* #143 (March 1977).

[17] *Daredevil* #5 (December 1964).

[18] *Daredevil Annual* #1 (1967), "Electro, and the Emissaries of Evil!"

[19] *Daredevil* #75 (April 1971).

[20] *Daredevil* #160 (September 1979).

[21] *Daredevil* #15 (April 1966).

[22] I thank Christine Hanefalk for this insight. Related to this, Hanefalk addresses the possible issues with Matt's efforts to "pass" as a blind person without enhanced senses in her post, "Faking It with Matt Murdock," at *The Other Murdock Papers* (https://www.theothermurdockpapers.com/2011/08/faking-it-with-matt-murdock). For a philosophical discussion of the more typical case of passing in reference to disability—written from the perspective of a legally blind person—see Adam Cureton, "Hiding a Disability and Passing as Non-Disabled," in Adam Cureton and Thomas E. Hill, Jr. (eds), *Disability in Practice: Attitudes, Policies, and Relationships*, Oxford: Oxford University Press, 2018, pp. 18–32.

[23] *Daredevil* #178 (January 1982).

[24] *Daredevil* #56 (September 1969).

[25] *Daredevil* #170 (May 1981).

[26] *Daredevil* #171 (June 1981). If you don't know who Fred and Ginger are, ask an old person—they'd probably appreciate the attention.

[27] I'm not a psychologist, so I will stop short of suggesting Matt might be bipolar. (For fascinating perspectives on Matt's psychology over the years and across media, see the essays in Travis Langley, ed., *Daredevil Psychology: The Devil You Know*, New York: Sterling, 2018).

[28] *Daredevil* #80 (September 1971), as are all the quotes and events to follow until noted otherwise. (If you want the read "can you dig it, hero-man" as a cry for help, that's your call.)

[29] Self-delusionary happiness is also a central theme of the run beginning with *Daredevil*, vol. 3, #1 (September 2011).

[30] *Daredevil* #81 (November 1971).

[31] *Daredevil* #82 (December 1971).

[32] Heather ended her life in *Daredevil* #220 (July 1985). Before she reappeared in *Daredevil* #227 (February 1986), Karen had not been seen in *Daredevil* since issue #86 (April 1972), except for one appearance in issue #138 (October 1976), in a story titled, appropriately, "Where Is Karen Page?"

[33] *Daredevil* #225 (December 1985), from which the events and quotes regarding the Vulture are also drawn.

[34] *Daredevil* #151 (March 1978). (Yes, he stammers even in his inner thoughts.)

[35] *Daredevil* #46 (November 1968).

[36] *Daredevil* #226 (January 1986).

[37] Thoughts like this also reflect self-obsession, a negative character trait that Matt exhibits during this period (but has been seen throughout his life in all his "woe is me" soliloquys). Others see it too: For example, when Matt's current love interest Glorianna complains about his silence and neglect, Foggy defends him, reminding her that Heather just died, to which Glorianna responds that Matt "seemed sorrier for himself than for her" (ibid.).

[38] *Daredevil* #227 (February 1986). Being Daredevil is the only source of joy left to him, as we see in the same issue when he anticipates his nocturnal adventures: "Something loose and wild flows through the city. I feel my pulse quicken, like a jungle drum. It's the night. I've always loved it. I grab the weightless bundle of cloth—the only part of my life worth living any more... the one relief I can give myself... when it all gets to be too much."

[39] Ibid. Nonetheless, this does not completely resolve his paranoia, as he still thinks to himself in the next issue, "Show me one single person in the world who hasn't betrayed me," and "all of them, working together. All of them out to get me," before remembering it was actually the Kingpin (*Daredevil* #228, March 1986).

[40] *Daredevil* #228 (March 1986).

[41] Ibid.

[42] Ibid. (As you can see, a *lot* happens in each of these issues—I'm only scratching the surface of the first few, so if you haven't read them, do yourself a favor and read them *now*.)

[43] *Daredevil* #230 (May 1986).

[44] This all happens in *Daredevil* #231-233 (June-August 1986). Captain America appears in the final issue and is particularly affected by the sight of a man, not so different from him, being manipulated by both the Kingpin and corrupt government officials to do their bidding. For more, see: https://thevirtuesofcaptainamerica.com/2021/04/23/daredevil-233-august-1986/.

[45] *Daredevil* #232 (July 1986).

[46] *Daredevil* #233 (August 1986).

Conclusion

The first two decades of the *Daredevil* comic, culminating in the work of Frank Miller and his amazing artistic collaborators, set the tone of the Man without Fear for years to come. Nearly every writer who has chronicled the adventures of Matt Murdock since then has either followed in Miller's footsteps or deliberately and consciously chosen a different path—either one proving Miller's influence on the character.

Without denying Frank Miller this influence, I hope I have shown in this book that, at least as far as the philosophical elements we have explored, Miller's Daredevil was not a reinvention but a consolidation and unification of what came before. As established by Stan Lee and reinforced by every writer that succeeded him, Matt Murdock has always been a man driven by both duty and desire, pushed to use his talents as a lawyer and superhero to do what is right while also trying to enjoy the joyful experiences life has to offer—two goals that often conflict, as we have seen, with sometimes catastrophic effects for Matt's mental well-being. Above all, Matt is a man of contrasts, which every creative team before and since has explored and depicted in their own way, but which Miller, chiefly in conjunction with Klaus Janson and David Mazzucchelli, showed more clearly than ever.[1]

If I've done my job in this book, I will not only have introduced you to many fascinating topics in philosophy, but I will also have given you new ways to think about Matt Murdock and Daredevil, whether that means looking at him differently than you did before, or just giving you a new way to express the opinions you already have. More specifically, I hope to have given readers a new appreciation for Daredevil's earliest stories, which are too often written off as silly or goofy, or of little value other than serving as a prelude to the cinematic brilliance of Miller and Associates. The Daredevil saga definitely reached a new level of maturity and stability with Miller's work, from which many more fantastic stories were spawned, but the early stories have their own value and charm as well. I encourage you to read them all, and read them often!

[1] For more on the visual elements of Frank Miller's storytelling in his first run of *Daredevil*, see Paul Young's penetrating study *Frank Miller's Daredevil and the Ends of Heroism* (New Brunswick, NJ: Rutgers University Press, 2016).

References

All comics cited in this book are listed below with their publication dates, writers, and artists (pencillers, inkers, and colorists); story titles are given only when there are multiple stories in the comic. Most issues are available digitally from Marvel.com or through Marvel Unlimited. (All issues of *Daredevil* are from volume 1 unless otherwise noted.)

Collections

Most of the earlier Daredevil comics have been collected in either Epic Collections, Marvel Masterworks, or dedicated collections, all listed below with relevant issues and referenced in the individual comics listings (with codes when easier). (The few comics I cite that are not from this period have their collections listed also when available.)

Daredevil Epic Collection volumes 1-7 and 17:

EC1: *The Man without Fear* (2016): *Daredevil* #1-21

EC2: *Mike Murdock Must Die!* (2018): *Daredevil* #22-41, Annual #1

EC3: *Brother, Take My Hand* (2017): *Daredevil* #42-63

EC4: *A Woman Called Widow* (2019): *Daredevil* #64-86

EC5: *Going Out West* (2022): *Daredevil* #87-107

EC6: *Watch Out for Bullseye* (2023): *Daredevil* #108-132

EC7: *The Concrete Jungle* (2024): *Daredevil* #133-154, Annual #4

EC17: *Into the Fire* (2023): *Daredevil: The Man Without Fear* #1-5.

Daredevil Marvel Masterworks volumes 1-18:

MM1 (1991): *Daredevil* #1-11

MM2 (2001): *Daredevil* #12-21

MM3 (2005): *Daredevil* #22-32, Annual #1

MM4 (2007): *Daredevil* #33-41

MM5 (2009): *Daredevil* #42-53

MM6 (2011): *Daredevil* #54-63

MM7 (2013): *Daredevil* #64-74

MM8 (2014): *Daredevil* #75-84

MM9 (2015): *Daredevil* #85-96

MM10 (2016): *Daredevil* #97-107

MM11 (2017): *Daredevil* #108-119

MM12 (2018): *Daredevil* #120-132

MM13 (2019): *Daredevil* #133-143, Annual #4

MM14 (2020): *Daredevil* #144-158

MM15 (2021): *Daredevil* #159-172

MM16 (2022): *Daredevil* #173-181

MM17 (2023): *Daredevil* #182-191

MM18 (2024): *Daredevil* #192-203

Frank Miller collections:

Daredevil by Frank Miller & Klaus Janson Vol. 1 (2010): *Daredevil* #158-161, 163-172

Daredevil by Frank Miller & Klaus Janson Vol. 2 (2010): *Daredevil* #173-184

Daredevil by Frank Miller & Klaus Janson Vol. 3 (2010): *Daredevil* #185-191

Daredevil by Miller & Janson Omnibus (2023): *Daredevil* #158-161, 163-191

Daredevil: Born Again (2012): *Daredevil* #227-233

Daredevil: The Man Without Fear (2010): *Daredevil: The Man Without Fear* #1-5

Other collections of early material:

Daredevil vs. Bullseye (2004): *Daredevil* #131-132, 146, 169, 181, 191

Daredevil: Love's Labors Lost (2002): *Daredevil* #215-217, 220-222, 225-226

Individual Comics

Amazing Spider-Man, vol. 1, #277, June 1986, "The Rules of the Game." Tom DeFalco (w), Ron Frenz, Bob Layton, and Bob Sharen (a).

Daredevil #1 (April 1964). Stan Lee (w), Bill Everett, Steve Ditko, Sol Brodsky, and unknown colorist (a). Collected in EC1 and MM1.

Daredevil #5 (December 1964). Stan Lee (w), Wally Wood and unknown colorist (a). Collected in EC1 and MM1.

Daredevil #7 (April 1965). Same as above.

Daredevil #9 (August 1965). Stan Lee (w), Wally Wood, Bob Powell, and unknown colorist (a). Collected in in EC1 and MM1.

Daredevil #15 (April 1966). Stan Lee (w), John Romita, Sr., Frank Giacoia, and unknown colorist (a). Collected in EC1 and MM2.

Daredevil #17 (June 1966). Same as above.

Daredevil #25 (February 1967). Stan Lee (w), Gene Colan, Frank Giacoia, and unknown colorist (a). Collected in EC2 and MM3.

Daredevil #28 (May 1967). Stan Lee (w), Gene Colan, Dick Ayers, and unknown colorist (a). Collected in EC2 and MM3.

Daredevil #29 (June 1967). Stan Lee (w), Gene Colan, John Tartaglione, and unknown colorist (a). Collected in EC2 and MM3.

Daredevil #41 (June 1968). Stan Lee (w), Gene Colan, John Tartaglione, and unknown colorist (a). Collected in EC2 and MM4.

Daredevil #42 (July 1968). Stan Lee (w), Gene Colan, Dan Adkins, and unknown colorist (a). Collected in EC3 and MM5.

Daredevil #43 (August 1968). Stan Lee (w), Gene Colan, Vince Colletta, and unknown colorist (a). Collected in EC3 and MM5.

Daredevil #46 (November 1968). Stan Lee (w), Gene Colan, George Klein, and unknown colorist (a). Collected in EC3 and MM5.

Daredevil #48 (January 1969). Stan Lee (w), Gene Colan, George Klein, and unknown colorist (a). Collected in EC3 and MM5.

Daredevil #49 (February 1969). Same as above.

Daredevil #53 (June 1969). Stan Lee and Roy Thomas (w), Gene Colan, George Klein, and unknown colorist (a). Collected in EC3 and MM5.

Daredevil #54 (July 1969). Roy Thomas (w), Gene Colan, George Klein, and unknown colorist (a). Collected in EC3 and MM6.

Daredevil #56 (September 1969). Roy Thomas (w), Gene Colan, Syd Shores, and unknown colorist (a). Collected in EC3 and MM6.

Daredevil #57 (October 1969). Same as above.

Daredevil #58 (November 1969). Same as above.

Daredevil #61 (February 1970). Same as above.

Daredevil #63 (April 1970). Same as above.

Daredevil #75 (April 1971). Gerry Conway (w), Gene Colan, Syd Shores, and unknown colorist (a). Collected in EC4 and MM8.

Daredevil #78 (July 1971). Gerry Conway (w), Gene Colan, Tom Palmer, and unknown colorist (a). Collected in EC4 and MM8.

Daredevil #80 (September 1971). Same as above.

Daredevil #81 (November 1971). Gerry Conway (w), Gene Colan, Jack Abel, and unknown colorist (a). Collected in EC4 and MM8.

Daredevil #82 (December 1971). Same as above.

Daredevil #86 (April 1972). Gerry Conway (w), Gene Colan, Tom Palmer, and unknown colorist (a). Collected in EC4 and MM9.

Daredevil #99 (May 1973). Steve Gerber (w), Sam Kweskin, Syd Shores, and Stan Goldberg (a). Collected in EC5 and MM10.

Daredevil #101 (July 1973). Steve Gerber (w), Rich Buckler, Frank Giacoia, and George Roussos (a). Collected in EC5 and MM10.

Daredevil #108 (March 1974). Steve Gerber (w), Bob Brown, Paul Gulacy, and Petra Goldberg (a). Collected in EC6 and MM11.

Daredevil #115 (November 1974). Steve Gerber (w), Bob Brown, Vince Colletta, and Petra Goldberg (a). Collected in EC6 and MM11.

Daredevil #116 (December 1974). Steve Gerber (w), Gene Colan, Vince Colletta, and Petra Goldberg (a). Collected in EC6 and MM11.

Daredevil #128 (December 1975). Marv Wolfman (w), Bob Brown, Klaus Janson, and Michele Wolfman (a). Collected in EC6 and MM12.

Daredevil #130 (February 1976). Same as above.

Daredevil #131 (March 1976). Same as above (also collected in *Daredevil vs. Bullseye*).

Daredevil #138 (October 1976). Marv Wolfman (w), John Byrne, Jim Mooney, and unknown colorist (a). Collected in EC7 and MM13.

Daredevil #139 (November 1976). Marv Wolfman (w), Sal Buscema, Jim Mooney, and Michele Wolfman (a). Collected in EC7 and MM13.

Daredevil #143 (March 1977). Marv Wolfman (w), Bob Brown, Keith Pollard, and Janice Cohen (a). Collected in EC7 and MM13.

Daredevil #146 (June 1977). Jim Shooter (w), Gil Kane, Jim Mooney, and Don Warfield (a). Collected in EC7, MM14, and *Daredevil vs. Bullseye.*

Daredevil #148 (September 1977). Jim Shooter (w), Gil Kane and Klaus Janson (a). Collected in EC7 and MM14.

Daredevil #150 (January 1978). Jim Shooter (w), Carmine Infantino and Klaus Janson (a). Collected in EC7 and MM14.

Daredevil #151 (March 1978). Roger McKensie, Jim Shooter, and Gil Kane (w), Gil Kane and Klaus Janson (a). Collected in EC7 and MM14.

Daredevil #159 (July 1979). Roger McKensie (w), Frank Miller, Klaus Janson, and Glynis Wein (a). Collected in MM15 and *Daredevil by Frank Miller & Klaus Janson Vol. 1 (and Omnibus).*

Daredevil #160 (September 1979). Same as above.

Daredevil #161 (November 1979). Same as above.

Daredevil #164 (May 1980). Same as above.

Daredevil #165 (July 1980). Roger McKensie and Frank Miller (w), Frank Miller, Klaus Janson, and Bob Sharen (a). Collected in MM15 and *Daredevil by Frank Miller & Klaus Janson Vol. 1 (and Omnibus).*

Daredevil #168 (January 1981). Frank Miller (w), Frank Miller, Klaus Janson, and D.R. Martin (a). Collected in MM15 and *Daredevil by Frank Miller & Klaus Janson Vol. 1 (and Omnibus).*

Daredevil #169 (March 1981). Frank Miller (w), Frank Miller, Klaus Janson, and Glynis Wein (a). Collected in MM15, *Daredevil vs. Bullseye,* and *Daredevil by Frank Miller & Klaus Janson Vol. 1 (and Omnibus).*

Daredevil #170 (May 1981). Frank Miller (w), Frank Miller, Klaus Janson, and Glynis Wein (a). Collected in MM15 and *Daredevil by Frank Miller & Klaus Janson Vol. 1 (and Omnibus).*

Daredevil #171 (June 1981). Same as above.

Daredevil #172 (July 1981). Same as above.

Daredevil #173 (August 1981). Frank Miller (w), Frank Miller, Klaus Janson, and Glynis Wein (a). Collected in MM16 and *Daredevil by Frank Miller & Klaus Janson Vol. 2 (and Omnibus).*

Daredevil #175 (October 1981). Frank Miller (w), Frank Miller, Klaus Janson, Christie Scheele, and Bob Sharen (a). Collected in MM16 and *Daredevil by Frank Miller & Klaus Janson Vol. 2 (and Omnibus).*

Daredevil #177 (December 1981). Frank Miller (w), Frank Miller, Klaus Janson, and Glynis Wein (a). Collected in MM16 and *Daredevil by Frank Miller & Klaus Janson Vol. 2 (and Omnibus).*

Daredevil #178 (January 1982). Same as above.

Daredevil #181 (April 1982). Frank Miller (w), Frank Miller and Klaus Janson (a). Collected in MM16, *Daredevil vs. Bullseye,* and *Daredevil by Frank Miller & Klaus Janson Vol. 2 (and Omnibus).*

Daredevil #182 (May 1982). Frank Miller (w), Frank Miller and Klaus Janson (a). Collected in MM17 and *Daredevil by Frank Miller & Klaus Janson Vol. 2 (and Omnibus).*

Daredevil #183 (June 1982). Roger McKensie and Frank Miller (w), Frank Miller and Klaus Janson (a). Collected in MM17 and *Daredevil by Frank Miller & Klaus Janson Vol. 2 (and Omnibus).*

Daredevil #184 (July 1982). Frank Miller (w), Frank Miller and Klaus Janson (a). Collected in MM17 and *Daredevil by Frank Miller & Klaus Janson Vol. 2 (and Omnibus).*

Daredevil #190 (January 1983). Frank Miller (w), Frank Miller and Klaus Janson (a). Collected in MM17 and *Daredevil by Frank Miller & Klaus Janson Vol. 3 (and Omnibus).*

Daredevil #191 (February 1983). Frank Miller (w), Frank Miller, Terry Austin, and Lynn Varney (a). Collected in MM17, *Daredevil vs. Bullseye,* and *Daredevil by Frank Miller & Klaus Janson Vol. 3 (and Omnibus).*

Daredevil #196 (July 1983). Denny O'Neill (w), Klaus Janson and Christie Scheele (a). Collected in MM18.

Daredevil #199 (October 1983). Denny O'Neill (w), William Johnson, Danny Bulanadi, and Bob Sharen (a). Collected in MM18.

Daredevil #200 (November 1983). Denny O'Neill (w), William Johnson, Danny Bulanadi, and Christie Scheele (a). Collected in MM18.

Daredevil #220 (July 1985). Denny O'Neil (w), David Mazzucchelli and Christie Scheele (a). Collected in *Daredevil: Love's Labors Lost.*

Daredevil #221 (August 1985). Same as above.

Daredevil #225 (December 1985). Denny O'Neil (w), David Mazzucchelli and Ken Feduniewicz (a). Collected in *Daredevil: Love's Labors Lost.*

Daredevil #226 (January 1986). Denny O'Neil and Frank Miller (w), David Mazzucchelli, Dennis Janko, and Christie Scheele (a). Collected in *Daredevil: Love's Labors Lost.*

Daredevil #227 (February 1986). Frank Miller (w), David Mazzucchelli and Christie Scheele (a). Collected in *Daredevil: Born Again.*

Daredevil #228 (March 1986). Frank Miller (w), David Mazzucchelli and Richmond Lewis (a). Collected in *Daredevil: Born Again.*

Daredevil #229 (April 1986). Frank Miller (w), David Mazzucchelli and Christie Scheele (a). Collected in *Daredevil: Born Again.*

Daredevil #230 (May 1986). Same as above.

Daredevil #231 (June 1986). Same as above.

Daredevil #232 (July 1986). Same as above.

Daredevil #233 (August 1986). Same as above.

Daredevil #347 (December 1995). J.M. DeMatteis (w), Ron Wagner, Bill Reinhold, and Christie Scheele (a). Collected in *Daredevil Epic Collection Vol. 20: Purgatory & Paradise* (2019).

Daredevil (vol. 3) #1 (September 2011). Mark Waid (w), Paolo Rivera, Marcos Martin, Joe Rivera, Javier Rodriguez, Muntsa Vicente (a). Collected in *Daredevil by Mark Waid, Vol. 1* (short volume, 2012) and *Daredevil by Mark Waid, Vol. 1* (long volume, 2013).

Daredevil (vol. 3) #24 (May 2013). Mark Waid (w), Chris Samnee, and Javier Rodriguez (a). Collected in *Daredevil by Mark Waid, Vol. 5* (short volume, 2014) and *Daredevil by Mark Waid, Vol. 3* (long volume, 2014).

Daredevil (vol. 4) #7 (October 2014). Mark Waid (w), Javier Rodriguez and Alvaro López (a). Collected in *Daredevil Vol. 2: West-Case Scenario* (short volume, 2015) and *Daredevil by Mark Waid & Chris Samnee, Vol. 4* (long volume, 2016).

Daredevil (vol. 5) #606 (October 2018). Charles Soule (w) and Phil Noto (a). Collected in *Daredevil: The Death of Daredevil* (2019).

Daredevil Annual #1 (1967), "Electro, and the Emissaries of Evil!" Stan Lee (w), Gene Colan, John Tartaglione, and unknown colorist (a). Collected in EC2 and MM3.

Daredevil: The Man Without Fear #1 (October 1993). Frank Miller (w), John Romita, Jr., Al Williamson, and Christie Scheele (a). Collected in *EC17* and *Daredevil: The Man Without Fear.*

Daredevil: The Man Without Fear #2 (November 1993). Same as above.

Daredevil: The Man Without Fear #3 (December 1993). Same as above.

Daredevil: The Man Without Fear #4 (January 1994). Same as above.

Daredevil: The Man Without Fear #5 (February 1994). Same as above.

About the Author

Mark D. White is Professor and Chair of the Department of Philosophy at the College of Staten Island/CUNY, where he teaches courses in philosophy, economics, and law. He is the author of ten books (including three in the *A Philosopher Reads...* series at Ockham Publishing), editor or co-editor of 19 more, and has written over 70 academic journal articles and book chapters in the intersections between economics, philosophy, and law.

You can find more information about Mark's books, articles, and blogs at www.profmdwhite.com, and you can follow him on most social media platforms under @profmdwhite.

www.ingramcontent.com/pod-product-compliance
Lightning Source LLC
Chambersburg PA
CBHW061831040426
42447CB00012B/2913